4th International Conference on Mechanical, Automotive and Aerospace Engineering (ICMAAE'18)

IOP Conference Series: Materials Science and Engineering
Volume 488

Kuala Lumpur, Malaysia
19 - 20 September 2018

ISBN: 978-1-5108-8483-0
ISSN: 1757-8981

Printed from e-media with permission by:

Curran Associates, Inc.
57 Morehouse Lane
Red Hook, NY 12571

Some format issues inherent in the e-media version may also appear in this print version.

Printed by Curran Associates, Inc. (2019)

For permission requests, please contact the Institute of Physics at the address below.

Institute of Physics
Dirac House, Temple Back
Bristol BS1 6BE UK

Phone: 44 1 17 929 7481
Fax: 44 1 17 920 0979

techtracking@iop.org

Additional copies of this publication are available from:

Curran Associates, Inc.
57 Morehouse Lane
Red Hook, NY 12571 USA
Phone: 845-758-0400
Fax: 845-758-2633
Email: curran@proceedings.com
Web: www.proceedings.com

Table of contents

Volume 488

4th International Conference on Mechanical, Automotive and Aerospace Engineering

19–20 September 2018, Kuala Lumpur, Malaysia

Accepted papers received: 16 January 2019
Published online: 20 March 2019

Preface

Papers

FOREWORD FROM THE 4th ICMAAE'18 CHAIRMAN

PROF. DR. MOHIUDDIN
Chairman
International Conference on Mechanical, Automotive and Aerospace Engineering 2018

Assalamu'alaikum Warahmatullahi Wabarakatuh,

All praise to ALLAH SWT. It is my pleasure to extend a warm welcome to all delegates attending the 4th International Conference on Mechanical, Automotive and Aerospace Engineering (ICMAAE'18) at the International Islamic University Malaysia in Kuala Lumpur. The conference organizers have put together excellent scientific programs that encompass both the latest research in mechanical engineering and provide an opportunity to renew old friendships and make new acquaintances.

Mechanical, Automotive and Aerospace engineering form the backbone of much of the industrialised world and play a vital part in steering the national goal of self-reliance and marching forward towards competitiveness in all areas of science and technology. It is hoped that the conference would provide a unique opportunity for academics, engineers and postgraduate students to meet, present and discuss the latest research developments, challenges and trends in mechanical, automotive and aerospace engineering.

The success of ICMAAE'18 depends completely on the effort, talent, and energy of researchers in the field of Mechanical, Automotive and Aerospace engineering who have submitted papers on a variety of topics. We are indeed glad at the favourable response received from the scientific community around the world. I would like to extend my thanks to the members of the organizing committee for their hard work in organising this excellent event. My thanks go to all the sponsors and all participants in making this conference a success.

I would like to take this opportunity to express my sincere gratitude and appreciation to all the reviewers who have helped in maintaining the high standard of the conference. It really gives me great pleasure to offer sincere thanks for your participation in ICMAAE'18. We hope that you will enjoy staying in Kuala Lumpur and take the advantage to see many beautiful sights in the city. We highly appreciate your presence and wish you every success in your research undertakings and look forward to seeing you at ICMAAE 2020.

Wassalam.

PROF. DR. AKM MOHIUDDIN

IOP Conf. Series: Materials Science and Engineering **488** (2019) 011002 doi:10.1088/1757-899X/488/1/011002

Peer review statement

All papers published in this volume of *IOP Conference Series: Materials Science and Engineering* have been peer reviewed through processes administered by the proceedings Editors. Reviews were conducted by expert referees to the professional and scientific standards expected of a proceedings journal published by IOP Publishing.

Experimental investigation of concentration yields of liquid phase exfoliated graphene in organic solvent media

A. Arifutzzaman[1], A. F. Ismail[2], I. I. Yaacob[1], M. Z. Alam[3] and A. A. Khan[1]
[1]Department of Manufacturing and Materials Engineering, Faculty of Engineering, International Islamic University Malaysia,
[2]Department of Mechanical Engineering, Faculty of Engineering, International Islamic University Malaysia, Malaysia
[3]Department of Biotechnology Engineering, Faculty of Engineering, International Islamic University Malaysia, Malaysia

E-mail: rahat_meengg@hotmail.com

Abstract. The main purpose of this work is to experimentally investigate the concentration yields of exfoliated Graphene in organic solvents with the variation of different centrifugation speeds (rpm). Graphene was effectively fabricated by one of the most convenient chemical processing approach liquid-phase exfoliation (LPE) of graphite in two different organic solvent N-Methyl-2-Pyrrolidone (NMP) and N, N-dimethylformamide (DMF). After exfoliation separated supernatant dispersions were preserved into the other tube as product. Similar procedure was repeated with same parameter and same graphite for both of the solvents. FESEM analyses of the exfoliated Graphene samples revealed the presence of few layers of Graphene flakes stacking. Analysis of the Raman spectrum of the samples it was revealed that the precursor graphite flakes was converted to Graphene sheets. Concentrations of graphene suspensions in both solvents were measured from separately prepared standard graphs using Beer's law with the aid of UV-vis absorbance spectroscopy. Estimated exfoliated graphene concentration was acquired higher for lower centrifugation (rpm) for both of the solvent. It was also seen that for the same rpm concentration of Graphene was higher in NMP than that of in DMF.

1. Introduction

Graphene is one or several planar sheets of carbon atoms and these are united with sp^2 bond. It possesses unique physical and chemical properties such as, extremely high mobility of charge carriers, and thermo-electrical conductivity, enormously large surface area and high mechanical strength. It is now becoming the most widely used carbon based nanomaterial [1]. Different types of graphene have been fabricated using different techniques. Such as chemical vapour growth technique [2], bottom-up growth by wet ball-milling technique [3] and burning of Mg metal in solid CO_2 (dry ice) technique [4]. In these techniques macromolecules are suffered from the happening of side reactions. Moreover, containing of impurities with the end products, complicated and time consuming initial rig setup limits their implementations [5]. Besides, these techniques have very low final yielding of graphene. Exfoliation of expanded graphite has also been used to produce graphene [6] however; intercalated compounds limit the end application of this graphene. Current efforts have emphasized on the chemical processing of graphite, which shown the most anticipation to produce graphene suspension in organic solvents [7]. Significant quantity of graphene is prepared by one of the most convenient chemical processing approach LPE of graphite by organic solvents [8, 9, 10]. It has ability to produce enormous amount of graphene sheets. It is very simple, direct and one-step method to yield graphene by dispersing the available graphite powder in organic solvents [5]. Although some work has been conducted to produce Graphene using LPE technique in organic solvents, however, till now it has not been investigated the concentration yields of exfoliated Graphene in organic solvents with the variation of different centrifugation speeds (rpm).

1.1. Liquid-Phase Exfoliation (LPE) for Graphene Production

Microcrystalline graphite particles can exfoliate into separate graphene flakes by the interaction with solvent. To exfoliate the graphite into individual layers two types of mechanical force are needed to be activated. Application of shear force along the graphene surface is effectively responsible for the exfoliation of graphite [11]. The enthalpy of mixing is dependent on the balance of surface energies of Graphene and the solvent [12]. Graphite surface energy is defined as the energy per unit area minimum required to overcome the van der Waals forces when peeling two sheets apart. A successful occurring of exfoliation depends on the small net energetic cost of the whole process [8]. The matching of the surface energies of Graphene and the solvent is one of the most prominent criterions for a successful exfoliation process. During the exfoliation process, if the surface energy of the solvent closer to the surface energy of Graphene then the enthalpy of mixing will be smaller and the degree of exfoliation will be higher [13]. The solvents NMP and DMF are properly fulfilling the criteria of mothering of surface energy with the Graphene. Thus, these solvents are intensely recommended for the successful exfoliation to graphene sheets [14].

Sonication parameters such as sonication time have been used from 30 minutes to 1000 hours. Intense tip sonication is relatively ineffective for the exfoliation graphene from graphite [15]. Sonication generates the shear forces and creates cavitation in the solvent liquid [11]. Due to the fluctuations of pressure small micro-sized voids or bubbles is created, grown and simultaneously those are collapsed in the liquids. It induces shock waves that surround the graphite and persuade exfoliation into graphene sheets [16]. Sedimentation-based separation (SBS) approach could divide the suspended flakes on the basis of degree of sedimentation due to the effect of centrifugal force on the flakes. Centrifugation of 500 rpm is sufficient to remove all the existing graphitic crystallites [8] from the dispersed flakes [15]. Thus controlled centrifugation is the potential mechanism to get the controlled concentration of graphene suspension from sonication assisted exfoliation dispersion. Suitable range of centrifugation speeds is from 4000 to 500 rpm for the analysis of sonication assisted exfoliated graphene concentration [15].

2. Materials and Equipment Used

Graphene flakes were fabricated using an easy and effective LPE technique. In this technique graphite flakes of molecular weight (MW), 12.011C (g/mol) were used in all experiments as the precursor material. They were purchased from Fisher Scientific (M) Sdn. Bhd (product code: 385030010, Lot: A0335167, CAS: 7782-42-5). Organic solvent NMP (C5H9NO, MW: 99.13) 99.9+% and DMF (C3H7NO, MW: 73.09) 99.8+%, were also supplied by Fisher Scientific (M) Sdn. Bhd (NMP product code: 390680010, Lot: A0333185, CAS: 872-50-4 and DMF product code: 423640010, Lot: A0340988, CAS: 68-12-2). Chemical was used as provided. Bath sonication was performed using a BRANSONIC (ultrasonic cleaner), model- 3510E-MTH, 100 W, 42 kHz. The supernatant was separated from sediment by removing the top ~80% of the dispersion by pipette. Centrifugation was carried out using an Eppendorf Centrifuge 5804 for 45 min in all cases. After exfoliation morphology of the exfoliated Graphene flakes were investigated using field emission scanning electron microscopy (FESEM) (Model JEOL JSM-6700F). Besides, exfoliated Graphene were also characterized by Raman Spectroscopy. It was performed on the prepared films on silicon wafer using an InVia Reflex, Renishaw, UK Raman Spectroscope. Concentrations of Graphene suspensions in solvents were estimated from the standard graphs prepared using Beer's law with the aid of ultra-violate visible (UV-vis) absorbance spectroscopy (Thermo Scientific, Multiskan GO, Version-1.00.40. 96-grid transparent micro-plates).

2.1. Experimental Procedures

First, 200 mg of graphite flakes was weighted using analytic balance and 200 ml of NMP was taken by a measuring cylinder. Then, NMP was poured on the graphite flakes in a beaker. The mixture was then exfoliated by bath sonication for 3 hours. A controlled centrifugation process was used for the separation of flakes in LPE. After exfoliation process prepared dispersion was first centrifuged at 4000 rpm for 45 minutes. Larger flakes were tending to sediment at the bottom of the centrifuge tube. Supernatant of graphene in NMP was decanted from the sediment by take away of top ~ 80 % of the dispersion by pipette without disturbing the sedimentation. Separated supernatant dispersion was then preserved into the other tube for characterization. Then 16 ml of NMP was added into the sedimentation and re-dispersed by a mild sonication of 15 minutes and centrifuged with 3000 rpm for 45 minutes. Supernatant of graphene in NMP was decanted again from sediment by take away of top ~ 80 % of the dispersion. This procedure was further repeated for three times, centrifuging the re-dispersed sediment at 2000, 1000 and 500 rpm, after every time separating the supernatant.

After 500 rpm centrifugation, the procedure was stopped as it has been found 500 rpm for 45 minutes to be the minimum required [15] to remove un-exfoliated graphitic crystal. Centrifugation with 500 rpm is sufficient to remove all the graphitic crystallites from the dispersed flakes [15]. Each of the supernatant suspension was the products. Then the dispersed concentration of the suspension was measured by absorbance analysis. Same procedure was repeated for the solvent DMF with maintaining all the parameters and conditions same. During the experiment water level was maintained carefully at the exact position in the water bath sonicator. Two separate standard curves were produced by preparing the known suspension in NMP and DMF solvents. The Beer's law ($A= \alpha C_G l$) provides a linear relationship between an absorbance of light and the properties of a material through which light is passing. Absorption coefficients, α are identified [5, 8, 9] using the Beer's law with the web-length (λ) 660 nm. The value of absorption coefficients were varied for different solvents [8, 15].

3. Results and Discussion

3.1. FESEM

Typical FESEM images of exfoliated Graphene flakes are presented in Figure 1. Images were taken on the sample prepared from supernatant Graphene suspension separated by final centrifugation speed (500 rpm). Figure 1(a) and 1(b) are showing the representative samples in solvents NMP and DMF respectively for the centrifugation speed of 500 rpm. FESEM analyses of the exfoliated Graphene samples revealed the presence of few layers of Gr flakes stacking. It can be seen that most of the exfoliated flakes are very thin and Graphene layers are stacked one over another in an ordered manner [16].

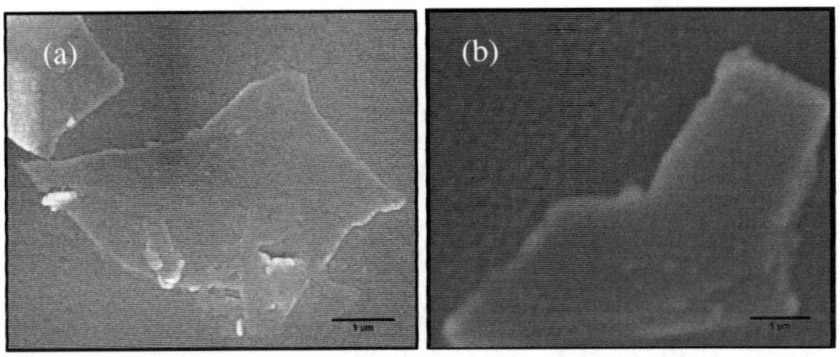

Figure 1. Representative FESEM images of exfoliated Graphene flakes of 500 rpm samples by solvent: (a) NMP and (b) DMF.

3.2. Raman Spectroscopy of Exfoliated Graphene

Representatives of Raman spectrum measured on the precursor graphite powder and samples of exfoliated Gr in DMF and NMP solvents for final centrifuge speed 500 rpm are shown in Figure 2.

The spectrum for the starting precursor powder is presented for the comparison purpose. Three characteristic peaks in the Raman spectra of graphitic carbon materials appear. In the measured Raman spectrum on precursor Graphite powder (PGP), G-band peak is found at ~ 1578.73 cm^{-1}, 2D-band at ~ 2648.96 cm^{-1} and D-band at ~1311 cm^{-1}. G band peak position for exfoliated Gr samples by NMP and DMF (supernatant of 500 rpm centrifugation) are shifted to ~ 1583.64cm^{-1} and ~1582.59 cm^{-1} respectively. It reveals the transformation of Gr form the graphite flakes by LPE. Increases in Raman shift (cm^{-1}) of G-band peak form the precursor graphite is the indication of the conversion of graphite to Gr sheets [17]. The sharp G-band peak appeared at around 1583 cm^{-1} in the spectrum of Gr, for the in-plane vibrational of sp^2 hybridized carbon atoms that comprise of the Gr sheet [17].

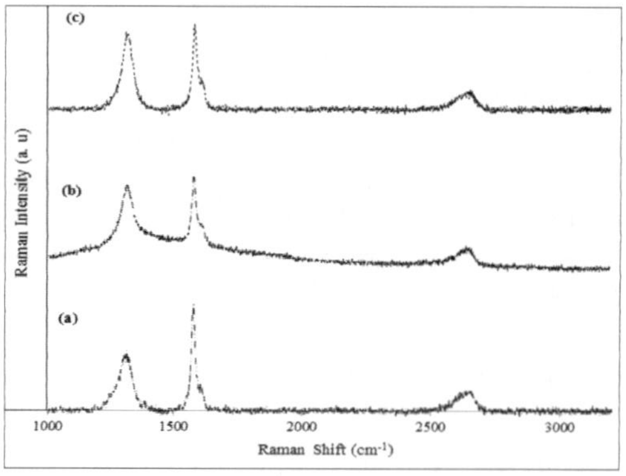

Figure 2. Raman spectra of (a) graphite powder (LPE precursor), exfoliated Graphene samples of 500 rpm centrifuge in solvent: (b) DMF and (c) NMP.

3.3. Physical Appearance of Exfoliated Graphene Suspensions
Digital photographs of supernatant graphene suspensions separated by varying centrifugation speeds (rpm) are shown in Figure 3. Although very slight levels of aggregation and sedimentation occurred within 30 days of centrifugation, the dispersions remain of high quality at least five months after preparation. Exfoliated Graphene suspensions in NMP are shown in Figure 3(a), 3(b), 3(c), 3(d) and 3(e) for the centrifugation speed of 4000, 3000, 2000, 1000 and 500 rmp respectively. Similarly exfoliated Graphene suspensions in DMF are shown in Figure 3(f), 3(g), 3(h), 3(i) and 3(j) for the same centrifugation speeds (rpm) respectively.

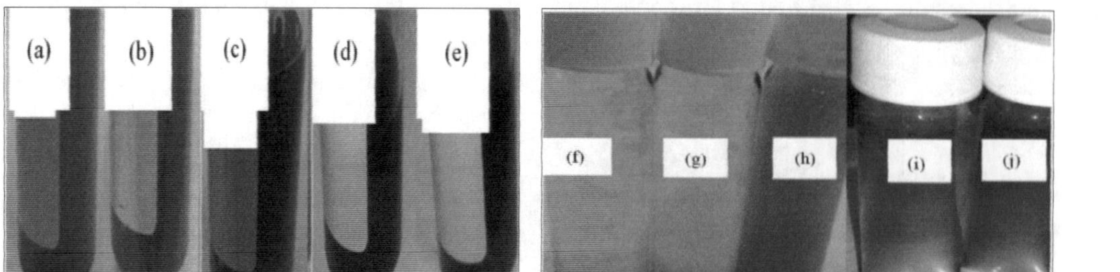

Figure 3. Dispersions of exfoliated graphene in NMP (a-e) and DMF (f-j) for the varying centrifuge speed from 4000 to 500 rpm respectively.

3.4. Concentration Estimation of Exfoliated Gr Suspension
Standard curves of the suspensions in NMP and DMF solvents are shown in Figure 4. Absorbance per cell length, (A/l) is obtained from these generated standard curves. In Figure 4a and 4b; the values of α is estimated 983.9 and 126.49 ml mg^{-1} m^{-1} for suspensions in NMP and DMF solvents respectively. Degree of correlations (R^2) between concentration, C_{Gr} and A/l are found 0.9984 and 0.9846 for NMP

and DMF suspensions respectively which are close to the unity. It indicates that instrument reading follow the Beer's law in readily with ideal case.

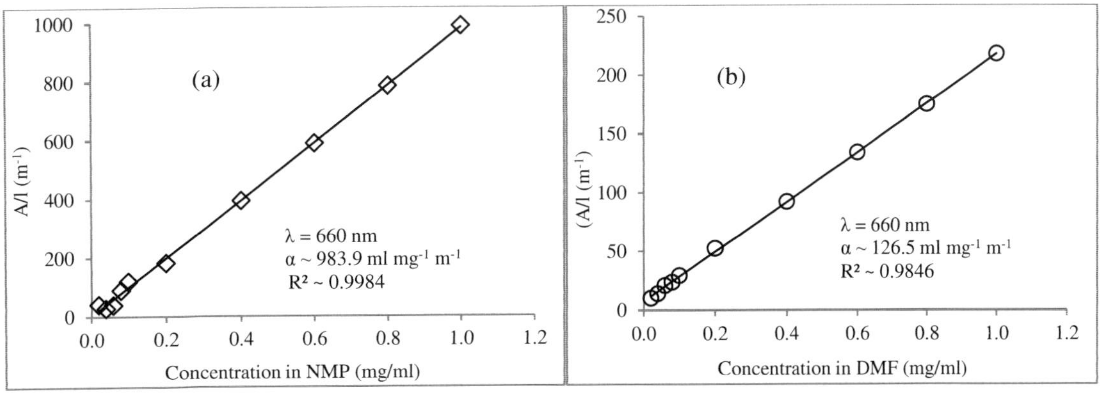

Figure 4. Standard curves of absorbance (λ = 660nm) per cell length, A/l (m^{-1}) verses concentration in two different solvents: (a) NMP and (b) DMF.

After sonication assisted exfoliation, graphene concentration, C_{Gr} of separated supernatant suspensions for varying centrifugation rates are estimated by the calculated value of absorbance per cell length, A/l (m^{-1}) of standard graphs once absorption coefficients, α is identified. C_{Gr} of supernatant graphene dispersions in NMP and DMF solvents verses centrifugation rate are plotted in Figure 5a and 5b respectively. Corresponding A/l (m^{-1}) is shown in the right axis of both graphs. It is seen that C_{Gr} is decreased monotonically with increasing centrifuge rate. It is agrees with work reported by Lotya et al., 2009 [9] and Khan, et al., 2010 [15]. The y-axis error bars represent the standard deviation calculated from measurements.

Concentration, C_{Gr} of 4000 rpm centrifuge supernatant solution (graphene suspension) in NMP (Figure 5a) is obtained 0.230 mg/ml. C_{Gr} of the collected supernatant suspensions of successively lower centrifugation speed of 3000, 2000, 1000 and 500 rpm are obtained 0.244, 0.268, 0.290 and 0.316 mg/ml respectively. While, C_{Gr} of exfoliated graphene suspensions (in figure 5b) in DMF are obtained 0.191, 0.222, 0.241, 0.260 and 0.313 mg/ml for the centrifugation rate 4000, 3000, 2000, 1000 and 500 rpm respectively.

Repeated exfoliation was conducted by maintaining exactly same procedure except for the different solvent DMF in lieu of using the solvent NMP. It is noted that C_{Gr} in NMP is higher than that of in DMF. Exfoliation of graphite is relying on the surface energy of solvents [8]. Where surface energy of solvents converted to enthalpy of mixing (per unit volume). For graphite, the surface energy is defined as the energy per unit area required to overcome the van der Waals forces when peeling two sheets apart [5, 8]. Surface energy of NMP solvent is perfectly matched [5, 8, 9] that of graphite were optimum exfoliation occurs. DMF surface energy is a bit lower than NMP resulting of higher exfoliation in NMP for the same sonication time from the same graphite precursor.

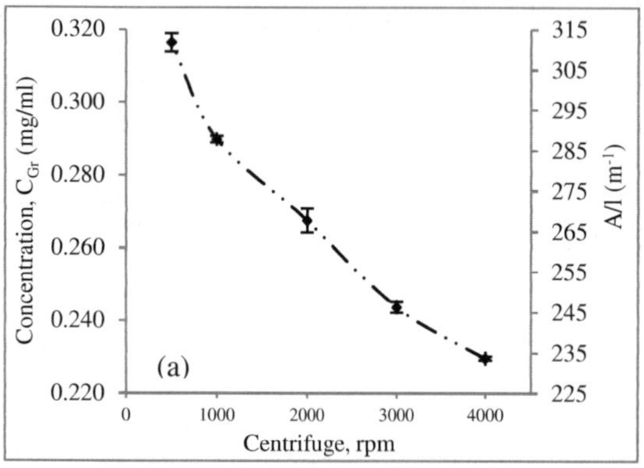

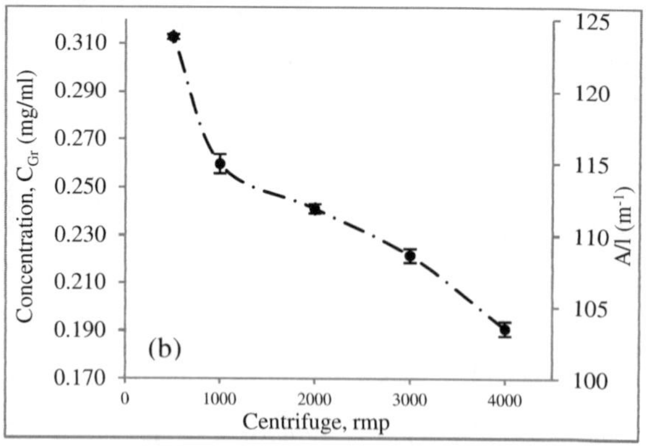

Figure 5. Exfoliated graphene concentrations as a function of centrifugation speed, rpm: (a) NMP and (b) DMF

4. Conclusions

Graphene flakes have been effectively fabricated using sonication assisted LPE separately in two different solvent NMP and DMF. Separate suspensions of exfoliated graphene were obtained using a systematic centrifugation scheme. FESEM analyses of the exfoliated Graphene samples revealed the presence of few layers of Graphene flakes stacking. Analysis of the Raman spectrum of the samples it was revealed that the precursor graphite flakes was converted to Graphene sheets. From the investigation of concentration of exfoliated Graphene suspensions, it is seen that the concentrations in lower centrifugation speed (rpm) is successively increased in both of the organic solvents NMP and DMF. In NMP concentrations of supernatant Graphene suspensions are obtained 0.230, 0.244, 0.268, 0.290 and 0.316 mg/ml for the successively lower centrifugation speed of 4000, 3000, 2000, 1000 and 500 rpm respectively. While in DMF, these are obtained 0.191, 0.222, 0.241, 0.260 and 0.313 mg/ml for same centrifugation rate (rpm) respectively. It is also perceived that, for the same centrifugation speed (rpm) concentration of Graphene suspension was higher in NMP than that of in DMF.

5. Acknowledgments

The authors courteously acknowledge the financial support provided by the Ministry of Higher Education (MOHE) under research grant ERGS 12-020-0020.

References

[1] Novoselov K, Geim A, Morozov S, Jiang D, Zhang Y, Dubonos S, Firsov A 2004 Electric field effect in atomically thin carbon films, *Science* vol 306(5696) PP 666-669
[2] Hu B, Ago H, Ito Y, Kawahara K, Tsuji M, Magome E, Mizuno S, 2012 Epitaxial growth of large-area single-layer graphene over Cu (111)/sapphire by atmospheric pressure CVD *Carbon* vol 50(1) pp 57-65
[3] Leon V, Quintana M, Herrero M. A, Fierro J L, de la Hoz A, Prato M and Vazquez E 2011 Few-layer graphenes from ball-milling of graphite with melamine *Chemical Communications*, vol 47(39) PP 10936-10938
[4] Arifutzzaman A, Yaacob I, Hawlader M A and Maleque M A 2015 Fabrication and Characterization of graphene from solid carbon dioxide *Advanced Materials Research* (vol 1115 pp 418-421) Trans Tech Publications
[5] Ciesielski A, Samori P 2014 Graphene via sonication assisted liquid-phase exfoliation *Chemical Society Reviews* vol 43(1) pp 381-398
[6] Zhou et al. 2014 Production of graphene by liquid-phase exfoliation of intercalated graphite. *International Journal of Electrochem Science* vol 9 pp 810-820

IOP Conf. Series: Materials Science and Engineering **488** (2019) 012001 doi:10.1088/1757-899X/488/1/012001

[7] Paredes J I, Villar-Rodil S, Solis-Fernandez P, Fernandez-Merino M J, Guardia L, Martinez-Alonso A and Tascon J M D 2012 Preparation, characterization and fundamental studies on graphenes by liquid-phase processing of graphite *Journal of Alloys and Compounds* vol 536 pp S450-S455

[8] Hernandez et al. 2008 High-yield production of graphene by liquid-phase exfoliation of graphite *Nature Nanotechnology* vol 3(9) pp 563-568

[9] Lotya et al. 2009 Liquid phase production of graphene by exfoliation of graphite in surfactant/water solutions *Journal of the American Chemical Society* vol 131(10) pp 3611-3620

[10] Coleman J N, 2012 Liquid exfoliation of defect-free graphene *Accounts of chemical research* vol *46*(1) pp 14-22

[11] Arao Y, Mizuno Y, Araki K and Kubouchi M 2016 Mass production of high-aspect-ratio few-layer-graphene by high-speed laminar flow *Carbon* vol 102 pp 330-338

[12] Hildebrand J H, Prausnitz J M and Scott R L 1970 Regular and related solutions: the solubility of gases, liquids, and solids *Van Nostrand Reinhold Company* New York

[13] Cui X, Zhang C, Hao R and Hou Y 2011 Liquid-phase exfoliation, functionalization and applications of graphene *Nanoscale* vol *3*(5) pp 2118-2126

[14] Compton O C and Nguyen S T 2010 Graphene oxide, highly reduced graphene oxide, and graphene: versatile building blocks for carbon-based materials *Small* vol 6(6) pp 711-723

[15] Khan U, Porwal H, O'Neill A, Nawaz K, May P and Coleman J N 2011 Solvent-exfoliated graphene at extremely high concentration *Langmuir* vol 27(15) pp 9077-9082

[16] Vadukumpully S, Paul J and Valiyaveettil S 2009 Cationic surfactant mediated exfoliation of graphite into graphene flakes *Carbon* vol 47(14) pp 3288-3294

[17] Hodkiewicz J 2010 Characterizing graphene with Raman spectroscopy *Thermo Scientific application note* pp 51946

IOP Conf. Series: Materials Science and Engineering **488** (2019) 012002 doi:10.1088/1757-899X/488/1/012002

Stability investigation of water based exfoliated graphene nanofluids

A. Arifutzzaman[1], A. F. Ismail[2], I. I. Yaacob[1], M. Z. Alam[3] and A. A. Khan[1]
[1]Department of Manufacturing and Materials Engineering, Faculty of Engineering, International Islamic University Malaysia,
[2]Department of Mechanical Engineering, Faculty of Engineering, International Islamic University Malaysia, Malaysia
[3]Department of Biotechnology Engineering, Faculty of Engineering, International Islamic University Malaysia, Malaysia

E-mail: rahat_meengg@hotmail.com

Abstract. Until now stability is a crucial technical challenge to prepare a homogeneous dispersion. The main purpose of this work is to experimentally investigate the suspension stability of the exfoliated Graphene (Gr) suspension in deionized water (DW) (Gr/DW). Five different samples were prepared by varying the exfoliated Gr volume fraction from 0.11 to 0.30 % in deionized water. For the observation of sedimentation photo capturing was used as a primary technique. Finally stability of the exfoliated Gr/DW nanofluids were monitored quantitatively with the aid of optical absorbance measurements using ultra violet-visible absorbance spectroscopy. The wavelength range of light was varied from 200 to 800 nm. Digital photographs of Gr/DW nanofluid samples after preparation of about 25 days shown homogeneous and visibly there was no sedimentation in the suspensions. The peak absorbance of Gr/DW nanofluids was appeared at the wavelength of 280 nm. For a certain volume fraction, absorbance was decreased monotonically with the increasing wavelength. The relative concentration of the Gr/DW samples was decreased due to slight agglomeration by the increasing concentration of precipitation with time. After 600 hours (~25 days), the lowest and maximum precipitation rate were appeared ~ 6 and ~ 10 % for the Gr/DW samples with the volume fraction of 0.11 and 0.30 % respectively. Different amount of exfoliated Gr loading affect the rate of sedimentations of exfoliated Gr/DW nanofluids.

1. Introduction

Stability is a crucial technical challenge to prepare a homogeneous dispersion because of the strong van der Waals interactions among the nanoparticles which always leading to the generation of aggregates. Sometimes sedimentation, clogging and aggregation phenomenon are the main cause for the stability of suspension which results in hampering of valuable properties of the suspension such as thermal conductivity and viscosity. There are some physical or chemical approaches such as surfactant addition, modification of particles surface or application of strong forces on the aggregated nanoparticles are implemented to achieve the stable nanofluids [1].

Even though, suspension stability of a nanofluid is an important concern for its application, there are inadequate studies on assessing the suspension stability of the nanofluids. Photo capturing is used as a primary technique to observe the sedimentation in nanofluids. Mehrali et al., (2014) [2] have investigated the Gr nanoparticle dispersed nanofluids by capturing the photos of the prepared different samples after 600 hours of preparation. Many other techniques such as, addition of surfactant [3], surface modification of Gr [4], chemical functionalization [5] and sonication [6] are commonly used for monitoring the suspension stability. However, these are not recommended techniques to monitor the stability of the nanofluids because of the consumption of longer time and qualitative accuracy of the stability results [7].

Ultra Violet-Visible (UV-Vis) spectrophotometer is the most widely used approach to estimate the rate or percentage of sedimentation of the nano-suspension [7]. It has been used to illustrate the colloidal stability of the nano-dispersions quantitatively.

At the wavelength from 200 to 800 nm, the UV-Vis spectrophotometer took the absorption by liquid and is used to analyze various dispersions in the fluid. Usually, stability of the suspension is determined by estimating the sediment concentration against the sedimentation time. However, for the high concentration (such as more than vol. 1%) this technique is shows some incompatibility for estimating dispersion stability of nanofluids [7, 8, 9].

2. Materials and Methods

Graphene nano-flakes were fabricated using a simple and direct technique sonication assisted liquid phase exfoliation (LPE) of graphite powder in organic solvent N-Methyl-2-Pyrrolidone (NMP). Preparation process and characterization are explained elsewhere [10]. After the dispersion of exfoliated Gr flakes in DW, it was bath sonicated using BRANSONIC ultrasonic cleaner for 10 minutes and then stirred about 1 hour using magnetic stirrer for the homogenization purpose. Gr/DW nanofluid samples were prepared by varying the Gr volume fraction in DW base fluid. Five different samples were prepared with the volume fraction of 0.10, 0.15, 0.20, 0.25 and 0.30 % of as produced Gr flakes.

2.1. Photograph Capturing

To see the sedimentation of produced Gr/DW nanofluid samples photo capturing was used as a primary technique. Some amounts of the suspension were to put aside after preparation to capture photos after a certain time. By comparing these photos of nanofluids, apparent sedimentation of suspension was observed.

2.2. Relative Stability Estimation by UV-vis Spectroscopy

The stability of the produced Gr/DW nanofluid samples were monitored quantitatively with the aid of optical absorbance measurements using UV-vis absorbance spectroscopy. The Beer-Lambert law ($A = \alpha l C$) provides a linear relationship between an absorbance of light and the properties of a material through which light is passing. A is the absorbance, α is the absorption coefficient (ml $mg^{-1}m^{-1}$), l (m) the distance that light travel through material and C is the concentration (mg ml^{-1}) of absorbing types in the material. Beer-Lambert law shows that at fixed volume fraction optical path and absorptivity, the absorbency is relative to the percentage amount of the particles inside the suspension.

At the first step, just after preparation each type of nanofluid samples were scanned in UV-Vis spectrometer (Thermo Scientific, Multiskan GO, Version-1.00.40. 96-grid transparent micro-plates) at room temperature while each measurement was repeated three times to attain a better accuracy. The wavelength range of light was varied from 200 to 800 nm. Peak absorbance was identified for every sample to a certain wavelength. Then, standard graph was fit for the each type of samples with sample concentrations (C_o) against their corresponding peak absorbance's as the concentration of suspension has a linear relation with absorbance. At the last step, relative stability of the as prepared nanofluids was monitored with increasing sedimentation time. In this purpose, UV-Vis spectrometer scanning was conducted on the supernatant of each sample in about 5 days (120 hours) interval for 25 days (600 hours). Supernatant concentration (C) of the samples was estimated for the corresponding absorbance. The relative concentrations (C/C_o) of the samples were plotted against time.

3. Results and Discussion

3.1. Physical Appearance of Gr/DW nanofluids

Figure 1 shows the physical appearance of the exfoliated Gr dispersed Gr/DW nanofluids samples. Digital photographs in Figure 1(a-e) show the representative of Gr/DW samples after preparation of about 25 days for the five different volume fractions. A homogeneous and uniform dispersion is acquired. It was apparent that there is no sedimentation in the suspensions.

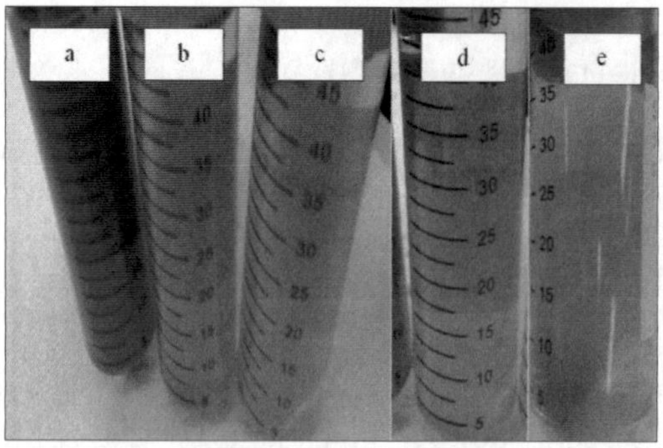

Figure 1. Dispersions of exfoliated Gr dispersion in DW. For the volume fraction of: (a) 0.30, (b) 0.25, (c) 0.20, (d) 0.15 and (e) 0.11 %.

3.2. Stability Investigation of Gr/DW Nanofluids with UV-vis spectroscopy

Figure 2(a) shows the absorbance of exfoliated Gr dispersed Gr/DW nanofluids samples for different amount loading with the variation of wavelengths. The peak absorbance of Gr/DW nanofluids is appeared at the wavelength of 280 nm. For a certain volume fraction the UV-vis spectrum of these Gr/DW nanofluids is featureless with a monotonic decrease in absorbance with increasing wavelength. Moreover, the absorbance of Gr/DW nanofluids decreases with the decreasing amount of Gr in DW; it should be known that the increasing amount of Gr will increase the absorbance that refers to the better nanofluid dispersion.

Figure 2(b) shows the relationships between volume fraction of Gr nanofiller and the corresponding peak absorbance obtained on the Gr/DW samples. There is a good linear relationship between the absorbance and the volume fraction of Gr is perceived, which satisfies Beer's law and indicates that Gr flakes were dispersed well in the DW base fluid [2] (Mehrali et al., 2014).

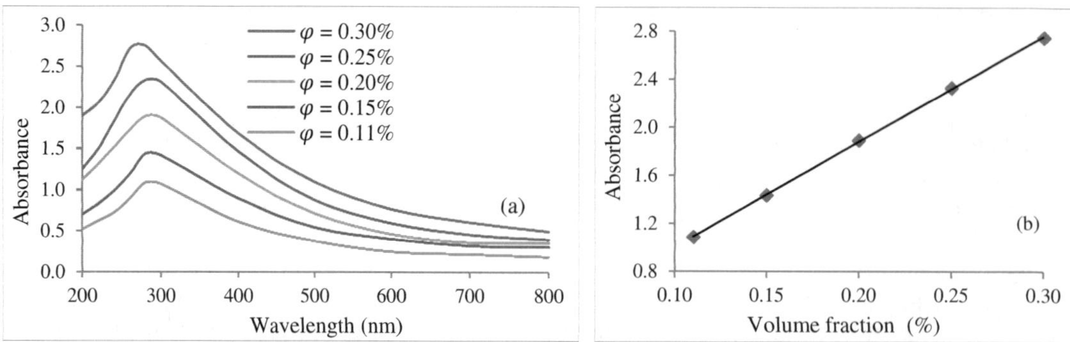

Figure 2. (a) UV-vis spectrum of Gr dispersed Gr/DW nanofluids samples at different volume fractions and wavelengths; (b) Linear relationship between light absorption and volume fraction of Gr in DW at wavelength of 280 nm.

Using the linear relationship (as shown in Figure 2(b)); Figure 3 shows the relative stability of Gr/DW nanofluids at different loadings as a function of sedimentation time. From the results, it is observed that, the relative concentration of the Gr/DW samples is decreased due to slight agglomeration by the increasing concentration of precipitation with time. Higher loading of Gr in DW leads to more sedimentation. Various forces such as gravitational force on the particles, electrostatic force and van der Waals forces between particles are effect in the nanofluid [2] (Mehrali et al., 2014). It is seen that, within 120 hours (~5 days) concentration decreased by 0.22, 1.46, 2.42, 3.98 and 4.26 % for the samples with the Gr loading of 0.11, 0.15, 0.20, 0.25 and 0.30 % respectively. The rate of sedimentation after 600 hours (~25 days) is different among these five samples as different amount of Gr loading are imposed. After this period of time, the lowest precipitation rate is appeared ~ 6 % for the Gr/DW sample with the volume fraction of 0.11%, while maximum precipitation rate is seemed ~ 10 % for the Gr/DW sample with the volume fraction of 0.30 %. These results show that variation of Gr volume fraction in the nanofluids affect the rate of sedimentation, which agree well with the results of previous studies by Mehrali et al. (2014) [2].

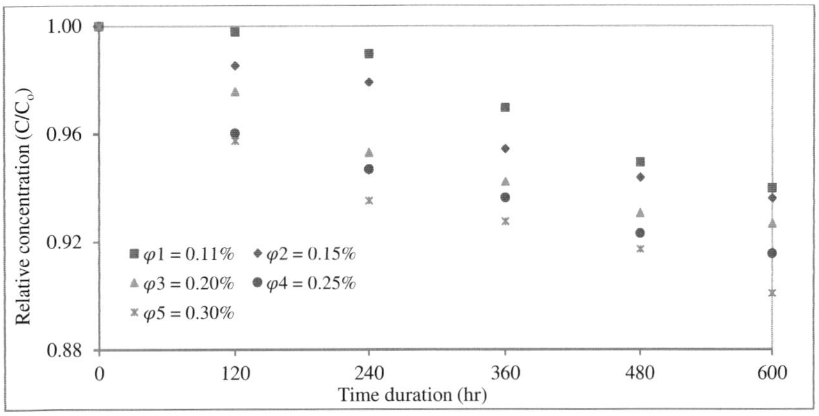

Figure 3. Relative supernatant concentrations of DW based exfoliated Gr dispersed nanofluids (Gr/DW) as a function of sedimentation time.

4. Conclusion

Stability of the exfoliated Gr dispersed Gr/DW nanofluids have been successfully investigated. It is found that that volume fraction loading has the effect on the rate of sedimentation as well as properties. Higher loading of Gr in DW leads to more sedimentation. After 600 hours (~25 days), the lowest precipitation rate is appeared ~ 6 % for the Gr/DW sample with the volume fraction of 0.11%, while maximum precipitation rate is seemed ~ 10 % for the Gr/DW sample with the volume fraction of 0.30 %.

5. Acknowledgments

Authors are grateful to the Research Management Centre (RMC), International Islamic University Malaysia (IIUM) for all kinds of support to accomplish this research work. The authors courteously acknowledge the financial support provided by the Ministry of Higher Education (MOHE) under research grant ERGS 12-020-0020.

References

[1] Park et al. 2002 Dispersion of single wall carbon nanotubes by in situ polymerization under sonication *Chemical physics letters* vol 364(3) pp 303-308

[2] Mehrali M, Sadeghinezhad E, Latibari S T, Kazi S N, Mehrali M, Zubir M N B M and Metselaar H S C 2014 Investigation of thermal conductivity and rheological properties of nanofluids containing graphene nanoplatelets *Nanoscale research letters* vol 9(1) pp 15

[3] Wen D and Ding Y 2004 Effective thermal conductivity of aqueous suspensions of carbon nanotubes (carbon nanotube nanofluids) *Journal of Thermophysics and Heat Transfer* vol 18(4) pp 481-485

[4] Johnson D W, Dobson B P and Coleman K S 2015 A manufacturing perspective on graphene dispersions *Current Opinion in Colloid & Interface Science* vol 20(5) pp 367-382

[5] Hao F, Fang D and Xu Z 2011 Mechanical and thermal transport properties of graphene with defects *Applied physics letters* vol 99(4) pp 041901

[6] Sarkar J, Ghosh P and Adil A 2015 A review on hybrid nanofluids: recent research, development and applications *Renewable and Sustainable Energy Reviews* vol 43 pp 164-177
[7] Ghadimi A, Saidur R and Metselaar H S C 2011 A review of nanofluid stability properties and characterization in stationary conditions *International Journal of Heat and Mass Transfer* vol 54(17-18) pp 4051-4068
[8] Velez E, Campillo GE, Morales G, Hincapie C, Osorio J, Arnache O, Uribe JI and Jaramillo F 2016 February Mercury removal in wastewater by iron oxide nanoparticles. In *Journal of Physics: Conference Series* (vol 687 No 1 pp 012050) IOP Publishing
[9] Liu Y, Jin W, Zhao Y, Zhang G, Zhang W 2017 Enhanced catalytic degradation of methylene blue by α-Fe_2O_3/graphene oxide via heterogeneous photo-Fenton reactions *Applied Catalysis B: Environmental* vol *206 pp* 642-652
[10] Arifutzzaman A, Yaacob I, Hawlader M A and Maleque M A 2015 Fabrication and Characterization of graphene from solid carbon dioxide *Advanced Materials Research* (vol 1115 pp 418-421) Trans Tech Publications

Hexapod robot for autonomous machining

Murshiduzzaman, Tanveer Saleh* and Md. Raisuddin Khan

Mechatronics Research Laboratory, Department of Mechatronics Engineering, International Islamic University Malaysia, Kuala Lumpur, 53100, Malaysia

*E-mail: tanveers@iium.edu.my

Abstract. Product miniaturization is a key aspect of manufacturing nowadays. Computer numerically controlled (CNC) machine tools are the major tools used in manufacturing industries for producing miniaturized products. However, CNC machines are still big, bulky and stationary. This research is an effort to develop a modular machine on a robotic platform which would be able to carry out machining operation. Hexapod robots are mobile, small and stable robots which are developed having a lot of variety. In this project a Hexapod robot was designed and fabricated for machining operation. The research scope for this project was limited to 1-D machining i.e. drilling operation. A suitable existing robot (hexapod) design was followed in this project. A drilling spindle was attached with the robot to carry out the machining operation. The robot was controlled using serial communication. A GUI (Graphical User Interface) was developed to control the Hexapod which had all the required algorithm inside. Machining operations were carried out with the prototype robot to test its performance. The highest accuracy was found using spindle speed of 2500RPM and velocity of 200mm/min. The repeatability found using these parameters were around 25micron. The positional accuracy of the robot movement was compared with that of an existing commercial micromachining system. The performance of the robot was found to be almost similar to that of the commercial machine.

Keywords: hexapod, legged mobile robot, miniature machining.

1. Introduction

CNC machine tools have been in use in the manufacturing sectors for more than 50 years [1]. It is used for micro fabrication. Recently, demand of micro-fabrication has increased tremendously due to product miniaturization. The modern researches about CNC machines usually focus on enhancing the performance of the CNC machining system. Some researches focus on the delivery of coolant at the place of machining [2]. Some focus on using multiple axis [3] while some other researches focus on incorporating and improving different types of operations with CNC machining [4]. However, for miniature product machining (feature size in mm range), conventional CNC machines are still used with significantly large work envelope and they occupy floor space permenantly. This is inefficient in terms of space utilization. Generally, CNC machines are bulky and very difficult to move from one place to another as they are not modular. As such, on site machining is not possible with current CNC machines.

There are many researches going on to optimize the functionality of hexapod robots. At the university of Duisberg, the department of Mechanics started to develop a hexapod named Tarry I in 1992, which followed the structure of stick insect (*Carausius morosus*) [5]. Later on there were some researches which considered different movement and structure of hexapod robots. Cockroach like structure was followed in a research in 1997 [6]. Many researchers have investigated kinematic and dynamic behavior of the hexapods [7]. There are various ways to design the motion patterns of the links of the hexapod

IOP Conf. Series: Materials Science and Engineering **488** (2019) 012003 doi:10.1088/1757-899X/488/1/012003

[8]. In 2012 some work on a hexapod robot called "HITCR II" by Jie Zhao, He Zhang, Yubin Liu, Jihong Yan, Xizhe Zang and Ziwei Zhou [9] . This robot was designed in a way so that it was capable of walking on an unstructured surface. A hexapod robot which was smaller than 9 cm in terms of length, width and height, was developed by Mayo Funatsu, Yushi Kawasaki, Soichiro Kawasaki and Koki Kikuchi in 2014. This robot had claws which assisted it to climb concrete wall [10].

In most researches, CNC machining and Hexapod robot had been usually two separate themes. In the research of hexapod robots, machining operations like CNC machining has not been focused much. Mostly the Hexapod robot researches focus on different designs, models and movements. In some recent researches, CNC machining is done by big size Hexapods [11]. This mobile hexapod give certain mobility and ease of movement of the CNC machining tool. Still the machine is somewhat heavy and bulky. This research can be thought of as the fusion of mini hexapod robots and CNC machining. The size of the machine tool was small because of using mini size Hexapods. Adjusting movement of the robot for rough terrain or obstacle was not covered in this research. In the current research, a suitable method for the walking mechanism of the hexapod was considered based on the literature review and more focus was given on the development of hexapod assisted micro/mini machining technology.

2. System Development

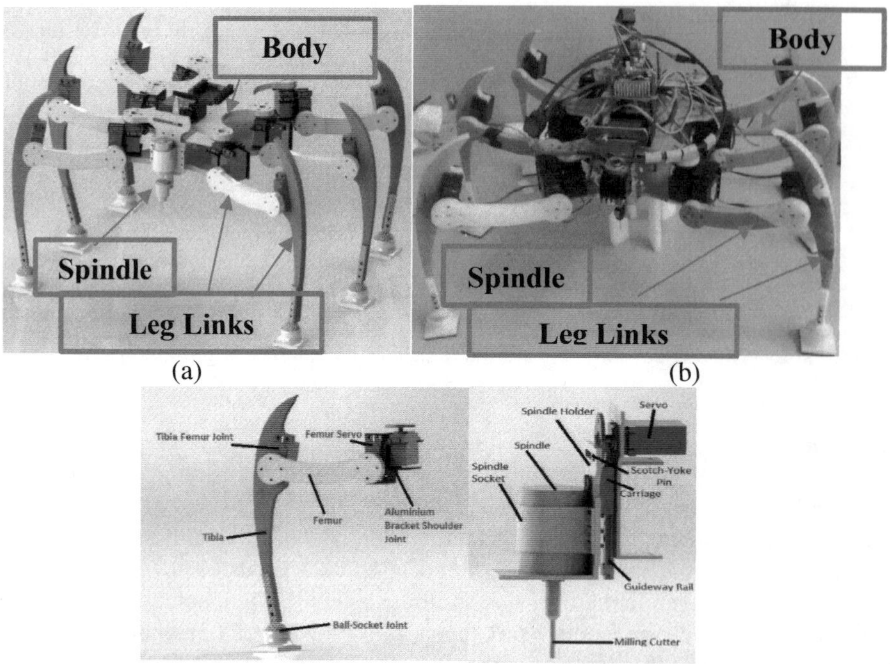

Figure 1. Complete Hexapod 3D design (a); and Actual Hexapod (b); 3D design of one leg with 3 servo and leg links (c) and Spindle design with scotch yoke mechanism (d)

The mechanical structure of the robot used in this research was a hexapod robot. The robot had six legs and each legs had two links and three motors. The design of this hexapod was partially done by following "Phoenix Hexapod" of Lynxmotion [12]. Overall 3 dimensional view of the complete hexapod is shown in Fig 1(a). The image of the fabricated robot in this research is also shown in Fig 1(b). The design of "Phoenix Hexapod" of Lynxmotion was followed while designing the robot in this research. But the Phoenix Hexapod of Lynxmotion was made of aluminium whereas the hexapod of this research was made of perspex. For this reason, some alteration was done in designing the links. The thickness was increased and some hollow features were avoided to provide more strength.

The main section of the robot is the body frame which acts as the anchor point for legs and co-axial (coxa) elements. The power supply is also housed internally in the body frame. The electronics which controls the movements of the robot are mounted on top of the body. The shape of the body frame is developed using two plates (top and bottom). Both of these plates are 8 mm thick.

Each leg has three servo motors. As shown in Fig 1(c), two motors are mounted in aluminum bracket to make the shoulder joint. There are two main links in each leg. They are femur and tibia. At neutral standing position femur would be horizontal. Tibia is connected with ball-socket joint. The ball-socket joint works like a feet for the hexapod robot. The ball-socket joint is attached with the tibia by using nut and bolt. The femur and tibia was at first designed with 3d design software solidworks. Then laser cutting was done. After finishing laser cutting, thread was formed by using tapping in the servo horn attaching holes. The ball-socket joint was 3D printed. Two parts were 3D printed separately and then they were attached by press fit technique. There is a rubber pad attached at the bottom of the ball-socket joint. This rubber pad increases the friction at the feet which gives good stability and less slip while machining or moving.

The spindle motor for this research was a 24-watt DC motor. It was clamped with the aluminium bracket which was attached on top of the carriage of linear guide way. Skotch-yoke mechanism was used to convert the rotational motion into Z-axis motion. The linear guide-way was used to make this motion smooth. Following proper installation, the machining spindle is shown in Fig 1(d). The cutting speed generated by the spindle can be calculated by the following equation.

Cutting speed = $(RPM * 3.14 * D)$ mm/min [13]

The diameter of the used cutting tool was, D = 0.9 mm

Using spindle speed of RPM 2500, the cutting speed can be calculated 7065.26 mm/min,

And using spindle speed of RPM 3500, the cutting speed can be calculated 9890 mm/min

The main electric components used in fabricating this hexapod robot are discussed in this section. One of the major components of the hexapod robot is a servo driver called SSC-32u. It can control up to 32 servo at once. Serial communication was done using USB cable. And also the battery used was Lithium-Polymar (LiPo) battery which provided 11.1 volt. But the servos and the servo driver required 6V to 7.4 volt. So a voltage regulator was needed. So a 12 ampere Voltage regulator was bought. The voltage regulator took 11.1 volt as input and gave 6.15volt output which was passed to other electrical components like SSC 32u.The spindle was controlled by md10c board shown in Fig 2. The signal was sent to md10c from SSC 32u servo driver.

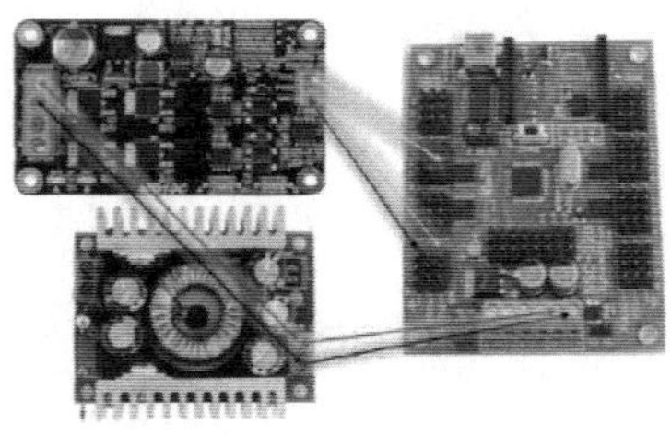

Figure 2. Circuit diagram of the Hexapod robot

A circuit diagram is shown in the Fig 2 which represents the connection of the hexapod. Other than this shown connection, there are 19 servos attached with different pins of the servo driver shown in the left side. The blue lines are the ground connection. The red connections are the positive connections.

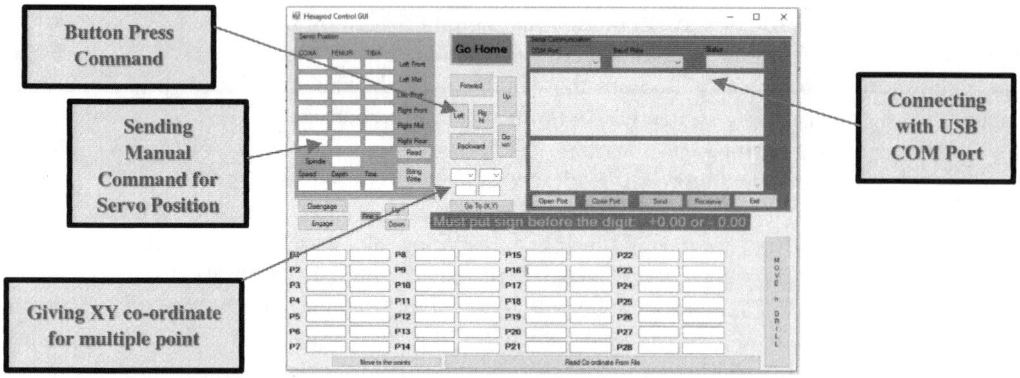

Figure 3. Graphical User Interface (GUI) of the Hexapod Robot

In the Fig 3 the GUI of the hexapod robot is shown. This gives an overview of the GUI. At first, the SSC-32u driver of the hexapod robot is connected via Universal Serial Bus (USB) with a computer which has the Hexapod robot control windows application. Next the specific port is selected where the hexapod is connected. If the connection is successful, then the progress bar of the GUI will show 100%. Otherwise it will ask the user to select the port details properly. After the port is connected then the user can start controlling the Hexapod. Now the User should click the big red button which is indicated as the "GO HOME" button. In this step the robot will become active. All the 18 motors will be activated and they will maintain a specific angle to give the robot home position. After this the robot will act as per the command is given. There are four different ways to proceed with the command from this stage.

3. Result and Discussion

After finishing the machining, the position and distance of the hole was measured. All the required measurement of the work piece was done by using Dino-Lite.

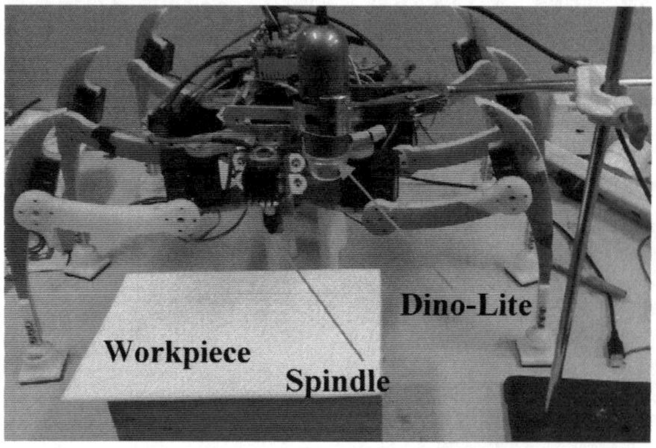

Figure 4. Experimental Setup

The experimental setup for doing characterization is shown in the Fig 4. The dino-lite was fixed with a holder to scan the drilled holes from above. The focus of dino-lite was adjusted in a way so that the holes were seen very clearly. The software Dinocapture 2.0 was used to see the captured image in computer. After fixing the focus, a virtual grid was generated using the software. Then calibration was done using a standard ruler to measure how much distance in the grid is equal to 1mm.

After calibration, the hole diameter was measured by using three-point circle measurement tool in the software. Magnification tool was used to detect the circle properly. Then the center to center distance from one circle to another was measured. This process was repeated for all the samples.

Table 1. Variable parameters

Point to point velocity		**Spindle Speed**	
V1	800mm/min	S1	3500 RPM
V2	400mm/min	S2	3000RPM
V3	200mm/min	S3	2500 RPM

The characterization data was recorded by varying spindle speed and point to point motion velocity. A 3x3 matrix was drilled every time when the parameters were changed. While doing the experiment, mainly two parameters were changed. The parameters are shown in a table 1. "V" indicates point to point moving velocity. And "S" indicates the spindle speed. V1 is the fastest velocity and V3 is slowest. S1 is the fastest spindle speed and S3 is slowest. Repeatability for each position was found by using the formula, Repeatability = Maximum error – minimum error. Using the repeatability data of the table, graphs were plotted to show the effect of the parameters.

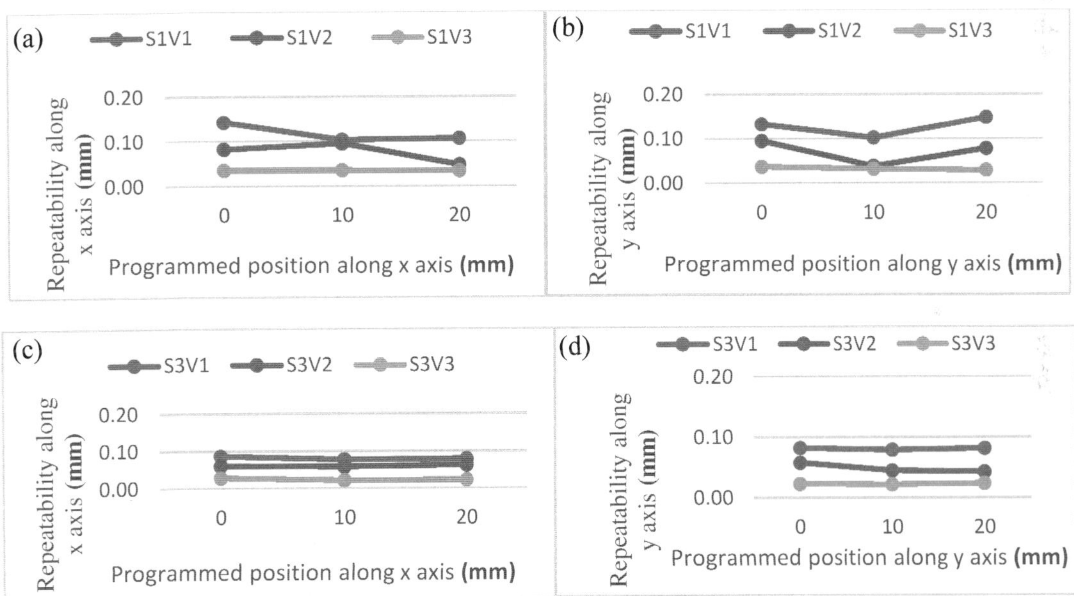

Figure 5. Graphical representation of repeatability found by using different parameters (a),(b)High spindle speed and (c),(d)Low spindle speed

Repeatability is considered good if the errors stay close to zero. After plotting the graph shown in Fig 5, it can be said that repeatability is best when the spindle speed and point to point velocity both are lowest. The explanation for this is that if the spindle speed is high, then the vibration caused by the spindle is high. This causes lower accuracy. And if the point to point velocity is too high, then there is chance that because of inertia while going to a point, the hexapod might have problem to go to desired point and might overshoot. But if the spindle speed and point to point velocity both are kept less, then these errors can be avoided and better result can be obtained.

4. Conclusion

In this research a new type of machine tool was fabricated which aims to be applied in such cases where moving the item to be machined is not a feasible option. The mobile robot fabricated in this research was a six legged hexapod robot. Since it has six legs, the stability is very good. The uniqueness of this research is the use of the control algorithm and the GUI developed in this research. Also the experiment focused on accuracy and repeatability for drilling operation which showed the capability of the machine tool in this research. The experiments were done varying drilling parameters. The parameters which were varied were spindle speed and point to point velocity. The effect of changing the parameters was observed and discussed. It was seen that the best result was found for the lowest spindle speed and lowest point to point velocity. After the experimentations, it can be said that hexapod robot and machining operation can be added together to make a mobile machining tool. However, some improvements can be applied in order to increase the ability of mobile machining. For milling operation, a better motion planning is required. For this reason, milling operation was not covered in this experiment. Doing milling operation using mobile robot like hexapod, can be an interesting research. Multiple robots can be used to do machining operation in order to finish the operation quicker. Parallel co-ordination can be used in this case. Using these recommendations some interesting research can be done in order to solve the problem of mobile machining platform.

References

[1] Thomas RA 2007 History of Numerical Control: *A History of the Role the General Electric Company Played in the Development of Numerical Control for Machine Tools*, 1943-88

[2] Robert Yellin DR 2017 Jul 27 *Apparatus and Method for Programmable Coolant Delivery in CNCMachines*.

[3] Gosselin C 2018 *Multi Axis CnC Manufacturing of Straight and Spiral Bevel Gears*. In Springer,Cham. p. 167–204.

[4] Tomal ANMA, Saleh T, Khan MR 2017 IOP Conference Series Mater Sci Eng . IOP Publishing; 260(1):012031 *Improvement of Dimensional Accuracy of 3-D PrintedParts using an Additive/Subtractive Based Hybrid Prototyping Approach.*

[5] Buschmann A 2000 *Home of Tarry I \& II: design of the walking machine Tarry II*. World Wide Web, http//www.tarry.de.

[6] Nelson GM, Quinn RD, Bachmann RJ, Flannigan WC, Ritzmann RE, Watson JT 1997 Proceeding of International Conference on Robot Automation *Design and simulation of a cockroach-like hexapod robot*.p 3–8

[7] Woering R. 2011 *Simulating the "First steps " of a walking hexapod robot*.;6–56.

[8] de Santos P, Garcia E, Estremera J 2007 Autonomous Robots 23(4) *Improving walking-robot performances by optimizing legdistribution*.p 247–58.

[9] Zhang H, Liu Y, Zhao J, Chen J, Yan J. 2014 *Development of a bionic hexapod robot for walking on unstructured terrain.*. 2014;11(2):176–87.

[10] Funatsu M, Kawasaki Y, Kawasaki S, Kikuchi K. 2014 J Bionic Eng *Development of cm -scale Wall ClimbingHexapod Robot with Claws*. p101–6.

[11] Tunc LT, Shaw J. 2016 International Journal on Advanced Manufacturing Technology 84(5–8) *Experimental study on investigation of dynamics of hexapod robot for mobilemachining*. 817–30

[12] Lynxmotion. *Phoenix 3DOF Hexapod* [Internet]. Available from: http://www.lynxmotion.com/p-947-phoenix-3dof-hexapod-no-electronics.aspx

[13] NS TOOL CO. LTD .*Calculation for Cutting Speed, Spindle Speed and Feed* |. [Internet]. Available from: https://www.ns-tool.com/en/technology/technical_data/cutting_speed/

IOP Conf. Series: Materials Science and Engineering **488** (2019) 012004 doi:10.1088/1757-899X/488/1/012004

Prototype of single degree of freedom optical resolver

Saiful Islam, Tanveer Saleh*, Marshad Latief, M.R.M Asyraf and Asan GA Muthalif

Smart Structures, Systems and Control Research Laboratory, Department of Mechatronics Engineering,
International Islamic University Malaysia, Kuala Lumpur, 53100, Malaysia

*E-mail: tanveers@iium.edu.my

Abstract. Optical resolvers are helpful for a wide variety of application, for example, automation and manufacturing industries. Different kinds of optical resolvers are available based on different material properties. In this paper, a prototype of a single Degree of Freedom optical resolver for measuring the angle of rotation is proposed. The device is based on the combination of the single wavelength (partially polarized) laser beam, a linear polarizer and a photodetector circuit is used for the rotational angle measurement. We present the result of sensitivity, accuracy and repeatability test to ensure the proper execution of single DOF optical resolver. The sensitivity of a single DOF optical resolver is around 0.467 V/degree and the standard deviation is 0.2 Volt. We obtained highly accurate and repeatable measurements according to the variation of the angle of the polarizer. Newly developed prototype optical resolver is cost effective, compact in size, no electromagnetic interference.

Keywords: polarization, prototype, optical resolver.

1. Introduction

Resolvers are generally utilized as a part of different fields of designing for angle position and speed location because of its robust structure, high reliability and responsive in the diverse environment [1-3]. Optical resolvers are helpful for a wide variety of application, for example, automation control system and the manufacturing industry. In recent times, the necessities for resolvers have been increasing with the development of the intelligent industry. Researchers have developed an optical resolver for measuring angle and position. Among them, Takashi Tokuda et al [4] developed a CMOS sensor to measure different incident polarization angle. Shiguang Li et al [5] introduced optical roll-angle sensor and the angle measurement based on the transmitted light passed through a polarizer analyzer, the transmitted optical intensity varied at every rotation of polarizer. Nicholas Albion et al [6] invented an optical angle sensor and explained two-stage and three-stage phase angle measuring sensor having an introduction of the period of a polarized light beam at 45° and 60° separately by rotating polarizer at a specific whole range of 180°. Clark et al. [7] showed a system to measure an angle of the aero-flexible deformation of a flying machine wing where polarized light was modulated to create two transporter beams, among them one was utilized as reference electrical signal and another as target electrical signal then the two were investigated by a phase indicator to decide the angle measurement. Huang et al.[8]introduced a new method of angle measurement system determined by the primary angle of incidence and polarization state of light. It was based on the polarization phase of light. Wijntjes et al.[9] invented noncontact optical angle encoder where polarizer was rotated synchronously to determine the angle of rotation. Previously, Butzer et al. [10] demonstrated a rotary motion detection system where linear polarizer and optical photodetector were used to identify the angle of rotation of a shaft. Thorburn et al.[11] proposed a position sensor utilizing polarizer and indicator which has the capacity to be used as a part of the push pin sensor inside a hard plate servo system. Hutchinson et al. [12] additionally built up an angular position transducer to acquire the rotation of a pole over a specific angle utilizing polarizer having coded tracks. Barnett et al. [13] explained a revolving sensor to identify the motion of a rotating shaft utilizing an optical linear polarizer. Subsequently, numerous methods have been implemented to develop optical resolver. In any case, they have impediments considering the intricate system architecture, large in size, complex circuitry procedure, accuracy, and precision. This paper proposes a prototype of optical resolver that can undoubtedly replacement of a traditional optical resolver. To acknowledge exceedingly perfect signals with the commercial resolvers, this resolver is intended to produce signals which particulars are the same as those of commercial resolvers. Additionally, a new type of optical resolver has no complex circuitry system, cost-effective, compact in size and highly reliable. In this paper, we present our experience in the fabrication and testing of the newly developed single DOF optical resolver.

2. Basic principle and Design

2.1. Structure

Structural model of single degree of freedom optical resolver mainly consists of three parts: a partially polarized laser source, a polarizer, and a photodetector.

Laser is used for providing light. It is emitting partially polarized light with constant wavelength. In figure 1, the partially polarized laser is fixed in the front side of the package. Partially polarized laser light is passed through a polarizer. According to Malus law [14], the partially polarized light will not reach zero for any angle of rotation of polarizer. Optical polarizer is fixed in between of laser and photodetector. The single wavelength partially polarized laser beam and a linear polarizer are utilized for the rotational angle measurement. The power of the transmitted partially polarized light was found differed by the angle variation of the polarizer.

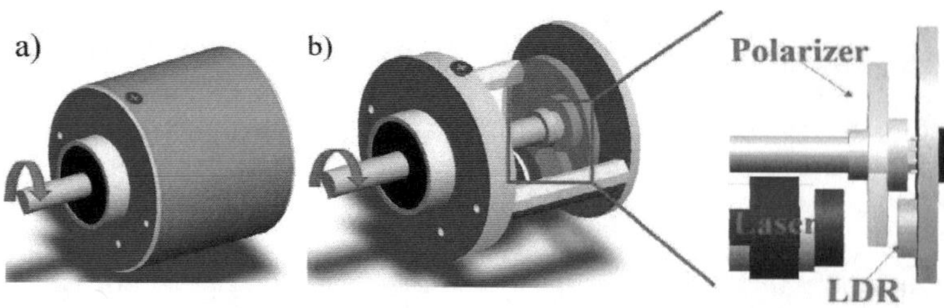

Figure 1. (a) & (b): Structural model of single degree of freedom optical resolver

Photodetector is situated in the ending part of the package as shown in figure 1. Transmitted light directed on to the photodetector circuit. Light dependent resistor is used to detect the light. It measure the voltage of transmitted light. It is situated in the ending part of the package.

IOP Conf. Series: Materials Science and Engineering **488** (2019) 012004 doi:10.1088/1757-899X/488/1/012004

2.2. Working principle

A single DOF optical resolver is including a laser source for providing partially polarized with constant wavelength light, an optical polarizer through which the light passes and a photodetector for measuring the voltage of transmitted polarized light as shown in figure 2. The polarizer is essential to set up a non-zero measurement of a partially polarized laser as opposed to utilizing completely polarized light. The principle of measuring an angle is varying the polarizer rotation angle.

From the derivation of malus's law, rana et al [14] explained input and output power of light beam. The equation of total input and output power is,

$$Pin = Pp + Pu \quad (1)$$

$$Pout = Pp \cos^2\theta + 0.5\, Pu \quad (2)$$

Here, Pp is the power of polarized light and Pu defined as the power of unpolarized light.

According to figure 3, an equation is developed from the ratio of Resistance of LDR and total output power, Pout. The equation is

$$Rldr = 0.43 Pout^{-0.425} \quad (3)$$

According to the voltage divider rule, Voltage across the known resistance is defined as the output voltage, Vout. According to the Figure 2,

$$Vout = \frac{R}{Rldr + R} \times Vin$$

$$\text{Or, } Vout = \frac{R}{R + 0.43 Pout^{-0.425}} \times Vin \quad (4)$$

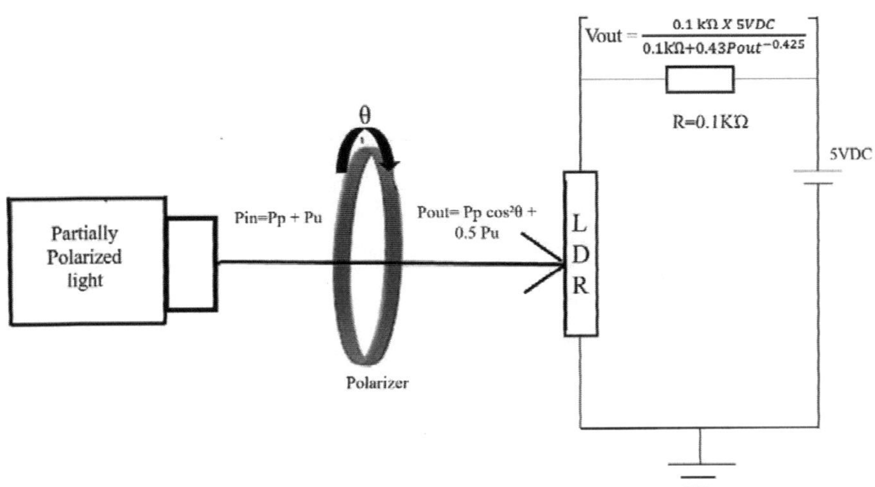

Figure 2. Schematic diagram of single DOF optical resolver.

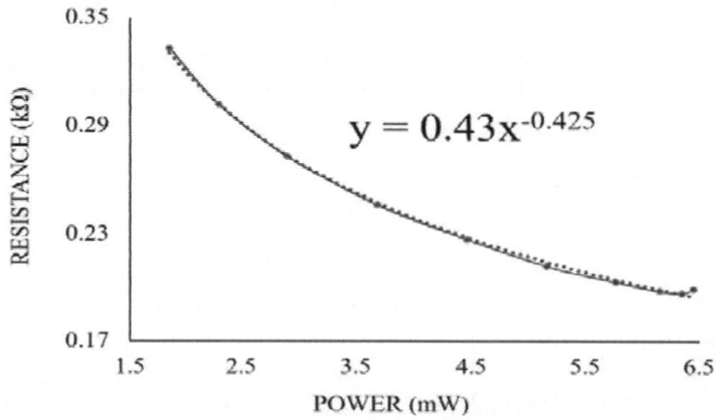

Figure 3. Graphical expression of Resistance of LDR Vs Power.

As per the standard, in part partially polarized laser light was passed through a polarizer. The polarizer was rotated from 0^0 to 360^0. After the laser light went through the polarizer, it was completely polarized. The measurement was taken by the photodetector. The output voltage of transmitted light was varied by the variation of revolution of the polarizer.

3. Experimental procedure

Numbers of examinations were led keeping in mind the end goal to legitimize the utilization of the linear polarizer and partially polarized light source to measure the rotational angle. Figure 4 shows a typical experimental setup that consists of a green laser source (partially polarized), an optical polarizer and 20 mm diameter photodetector. Photodetector circuit consists of 20 mm diameter LDR, 0.1KΩ Resistance and 5VDC as an input of the circuit. Photodetector circuit comprises of 20 mm diameter of as LDR, 0.1KΩ Resistance and 5VDC as a contribution of the circuit. At first, partially polarized laser source was turned on for a specific period to ensure it is balanced out before the experiment. From that point, photodetector and polarizer were adjusted on a similar arrangement as per the polarized beam. The measurements were taken by rotating polarizer at 10° angle shift and over the range of 360°. The transmitted light of the partially polarized light was captured by a photodetector. Strangely, the transmitted light changed by the angle variety of the polarizer at each of the interval over the entire range. Also, the increment and decrement of the voltage at each of the 90° interval was almost the same except in some degrees. We also performed an experiment using the different setting of the power from polarized source taking the measurement of the transmitted voltage over 0° to 360° and 360° to 0° to study on the repeatability.

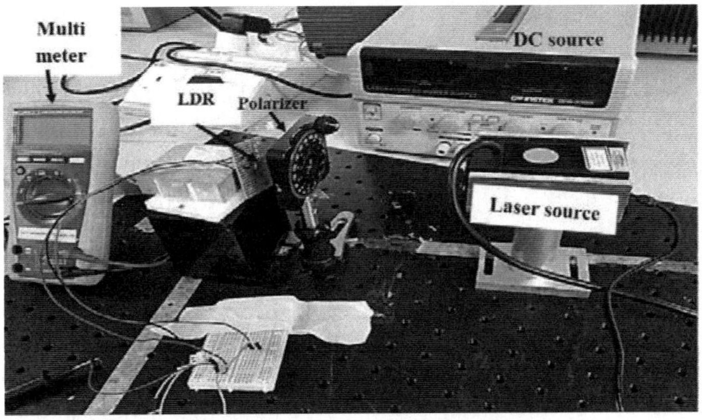

Figure 4. Schematic diagram of the whole assembly of the experiment.

4. Fabrication of a prototype

4.1. Thin film optical polarizer

Thin Film Optical polarizer need to cut to make a usable shape for the prototype. A servo-controlled 3-axis Micro-CNC machine (Mikrotools Ltd) is utilized to cut the thin film polarizer in a round shape. The outer diameter across of the polarizer is 30 mm. The measurement of the inner diameter is 5 mm.

4.2. Photo detector circuit

LDR (Light Dependent Resistor) is utilized to detect the transmitted light. Circuit depends on the hypothesis of voltage divider rule. 0.1kΩ Resistance, 12 mm diameter circular shape LDR and outside 5 VDC power source is utilized to initiate the circuit. Figure 5 demonstrates output and input wire originating from photodetector circuit.

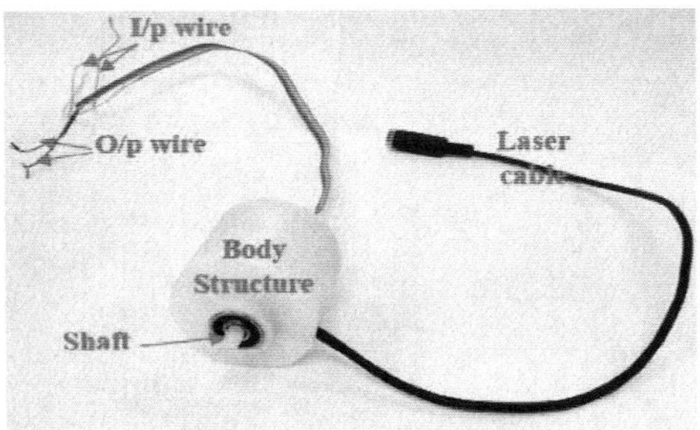

Figure 5. Prototype of Single Degree of Freedom Optical Resolver

4.3. Body structure

The shape of the optical resolver is round and hollow. Measurement is 35 mm. furthermore, the length is 40 mm. 3 D printer is utilized to make the assemblage of single degree of freedom optical resolver. The real structure and state of single DOF optical resolver appear in Figure 5.

5. Speed testing experiment of prototype

A test bench has been designed and built as shown in Fig 6. This setup utilized for testing the optical resolver driven by a variable speed stepper motor. A 5mm inner diameter coupler was utilized to couple the motor shaft and optical resolver shaft. That was the mechanical piece of the precision testing of the optical resolver. Arduino GrBl used to control the stepper motor's precise speed. 24 VDC provided from DC source to enact the controller. DC source provided 5VDC to active the photodetector circuit. The tests were directed with changing motor revolution speed. Curiously, the transmitted light shifted by the angle variety of the polarizer and revolution of the output signal was varied according to the variation of the angular speed of the stepper motor.

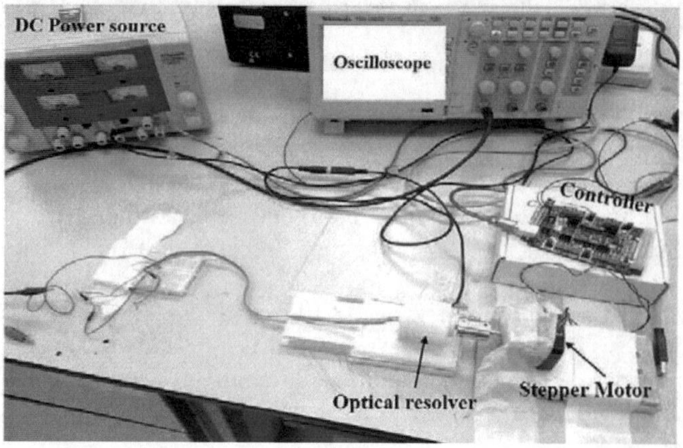

Figure 6. Experimental Step up of Accuracy Testing of Optical Resolver

6. Results and discussion

6.1. Experimental results

From the graphical representation of Figure 7, as the rotation angle of polarizer is increased, voltage increased. According to the operation principle polarizer rotated varying from 0^0 to 360^0 with an interval of 10^0. Higher sensitivity value is desirable for the good sensor application. In this experiment, the sensitivity achieved around 0.467 V/degree. In terms of Repeatability, standard deviation found around 0.2 Voltage. We acquired exceedingly precise and repeatable measurements as per the variety of the angle of the polarizer in each of the tests.

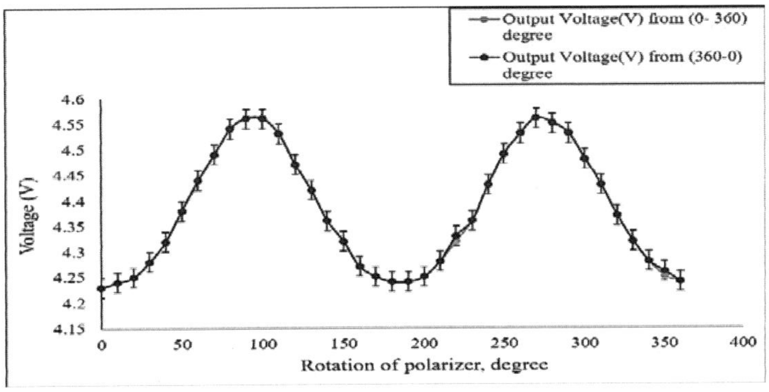

Figure7. The graph represents the experimental result of the whole assembly

6.2. Speed test result

Figure 8 demonstrates the aftereffect of the output voltage with a various angular rotation speed of the stepper motor. To look at the exactness of single DOF optical resolver. We did an examination to test the precision in various RPM. Figure 8(a) appears at 16 RPM optical resolver takes 2 sec to finish one cycle. At the point when the motor speed is increased, the time of the one revolution is decreased. At figure 8(d), it takes just 0.7 sec to finish full one cycle. Thus, we acquired from the analysis that, time is inversely proportional to speed which drove the probability of this prototype to make it as a good resolver. There is having bit fluctuations at a few degrees. The most probable explanation behind the overall fluctuations over the entire range is due to the sensitivity of the detector to its surrounding while capturing transmitted polarized power. This is on account of the ambient temperature, humidity of the environment and the impact of light normally interfere with the sensitivity of the photodetector. Then again, the contamination of the polarizer may have the impact to reduce the transmitted power from the real one.

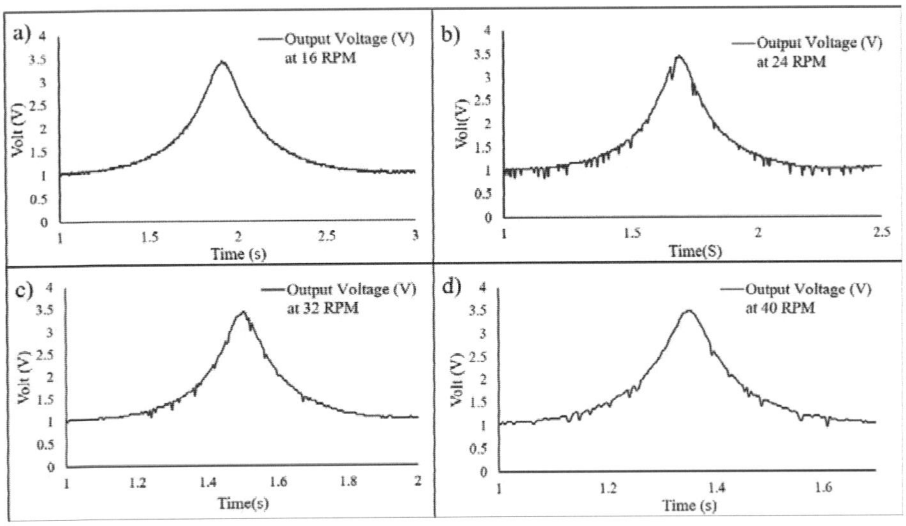

Figure 8. Experimental result obtained from different RPM

7. Conclusion

This paper proposed a new optical resolver compatible rotary position sensor. The optical resolver consists of the laser, polarizer and photo detector circuit. Transmitting light passes through the optical resolver and directed on to the photodetector. Based on the polarization of light, the operation principle was discussed. The operation was experimentally verified using a fabricated prototype. It proves its high sensitivity and accuracy. It is cost effective, compact in size. There is no electromagnetic interference as it depends on the optical standard of lasers.

References

[1] Masaki K, Kitazawa K,Mimura H, Nirei M, Tsuchimichi K 2000 Magnetic field analysis of a resolver with a skewed and eccentric rotor *Sensors and Actuators A: physical* **81** 297-300

[2] Benammar Mohieddine, Ben-Brahim Lazhar, Alhamadi Mohd A 2004 A novel resolver to 360 degrees; linearized converter *IEEE Sensors Journal* **4** 96-101

[3] Katakura Masayuki,Toda Asako, Takagi Yuichi, Suzuki Norihito, Kadoyama Takahide 2005 A 12-bits resolver-to-digital converter using complex *Symposium on VLSI Circuits Digest of Technical Papers* **15** 236-239

[4] Tokuda T., Yamada H., Sasagawa K. and Ohta J 2009 Polarization-Analyzing CMOS Image Sensor With Monolithically Embedded Polarizer for Microchemistry Systems *IEEE Transactions on* **3** 259-266

[5] Li S., Yang C., Zhang E. and Jin G 2003 Compact optical roll-angle sensor with large measurement range and high sensitivity *Optics Letters* **30** 242-244

[6] Albion N., Asmar R. F., Huggins R. W., Miller G. E., and Porter C. R 1995 Optical Angle Sensor Using Polarization Techniques *U.S. Washington DC: U.S. Patent and Trademark office* **5** 424-535

[7] Clark L. T 1987 Rotating Polarizer Angle System. *DC: U.S. Patent and Trademark Office* **4** 688-934

[8] Huang P. S., Kiyono S. and Kamada O 1992 Angle measurement based on the internal reflection effect a new method *Applied Optics* **31** 6047-6055

[9] Wijntjes Geert and Constantine Markos 2005 Non Contact Optical Polarization Angle Encoder *U.S. Patent* **10** 288-875

[10] Butzer D. C 1996 Polarization Based Optical Sensor Utilizing Total Internal Reflection *U.S. Washington, DC: U.S. Patent and Trademark Office* **5** 483-346

[11] Thorburn William G., Donald K. Mitchell and Bruce A. Horwitz 1998 Diffuse Surface Interference Position Sensor *U.S. Patent* **5** 783-752

[12] Hutchinson Paul L. and W. Gordon White 1993 Angular Position Sensor Using A Polarized Disc With Coded Tracks *U.S. Patent* **5** 235-177

[13] Barnett, Donald 2002 Optical Resolver And Method Of Use *U.S. Patent* **10** 095-542

[14] Rana Masud, Mrazib Masyraf, Saleh T., Muthalif Asan GA 2015 Development Of An Angle Sensor Using Optical Polarizer *ARPN Journal of Engineering and Applied Sciences* **10** 17416-17420

Aerodynamics of a circular planform wing

Suliman M. M. Ali[1], Ashraf A. Omar[2], Waleed Fekry faris[3], Ahmad Faris Ismail[4], J. S. Mohamed Ali[5]

Kulliyyah of Engineering, International Islamic University Malaysia

waleed@iium.edu.my

Abstract. This paper investigates the aerodynamics of a circular plan form wing at high Reynolds numbers numerically. Commercial software STAR-CCM+ has been used for the study. The numerical results were presented in terms of flow visualization, pressure over the wing, velocity contour and streamlines, lift and drag coefficients for angle of attacks ranging from 0° to 30° and Reynolds number equal to 510,000. The results showed that the distribution of the pressure is dominated by vortices behaviour at high angle of attack. The results also show that there is a large separation region located on the trailing edge of the elevator and the cockpit region. It is evident that this type of wing configurations plays a very important role in delaying or preventing the flow separation. The result agreed with available experimental data.

1. Introduction

Aircraft with low aspect ratio wings behave differently than high aspect ratio aircraft, such as the USAF B52 bomber and sail planes. A larger taper will lead to more unstable characteristics when compared to a low tapered wing with a high aspect ratio. Dating back to the early 1940's, several disk shaped aircraft were constructed and tested for both personal and military applications. Arthur Sack constructed a low aspect ratio circular planform aircraft [1]. During takeoff tests, it was noticed that the control surfaces were in a lower pressure area behind the circular wing.One of the most unusual aircraft ever designed for the U.S. Navy was the Chance Vought V-173, also known as the Zimmerman "Flying Pancake". It was a prototype "proof of concept" aircraft that lacked wings, instead relying on its flat circular body to provide the lifting surface [2]. As early as 1933, Charles Zimmerman had experimentally studied several airplanes with low aspect ratio wings and found a range of aspect ratios extending approximately from 0.75 to 1.50 [3].. He later determined from experiments that low aspect ratio wing designs were more efficient than conventional wings when the vortices from the wing were controlled with propellers [4].. One of the more interesting configurations tested was a military concept the Lenticular Reentry Vehicle or LRV . This was studied in the 1960's. Its design was considered because it created more lift than a standard wing, especially at low speeds [4]. . After testing several configurations in wind tunnels at subsonic speeds, it was found that the optimal LRV produced longitudinal stability, a positive pitching moment at zero angle of attack and a maximum lift to drag ratio of five. In the early 1970s the U.S. Navy commissioned a project in which the aerodynamic characteristics of a self-suspended Frisbee shaped flare was investigated [5].. Both spinning and non-spinning models were tested and it was found that spin had negligible effects on the aerodynamic forces and moments. Later, Stilley and Carstens [5].. a circular palnform wing was studied, it was concluded that low sect ratio aircraft has more stability [6].

In the early 1970s the U.S. Navy commissioned a project in which the aerodynamic characteristics of a self-suspended Frisbee shaped flare was investigated [7]. Both spinning and non-spinning models were tested and it was found that spin had negligible effects on the aerodynamic forces and moments.

In August 9, 2008 two Geobat models had been tested in Auburn University's wind tunnel test [8]. In this paper, the numerical study of the flow around IIUM designed circular wing remote controlled aerial vehicle was carried out to study the aerodynamic performance of the circular wing configuration.

2. Study model

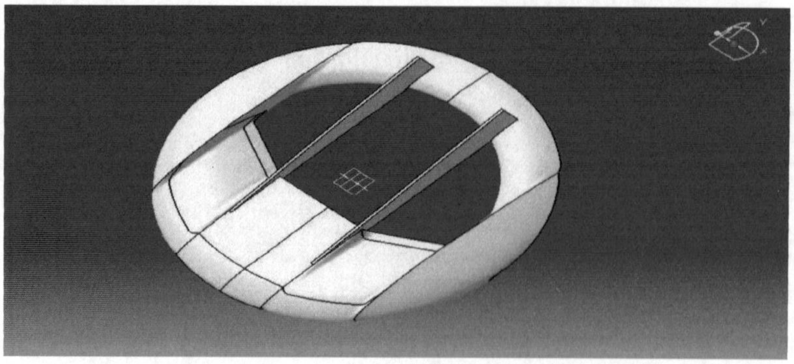

Figure 1. Circular wing geometry.

The studied models had a disk-shaped body of 0.72 m outer diameter with a central opening. The model can be described as a joined wing with a circular arc rearward swept front wing, a forward-swept rear wing with a circular trailing edge and two connecting wing tips, thus creating a 360 degree circular planform. The control surfaces included flaps, rear ailerons, a large elevator and two rudders. Each section of the wing was contoured with NACA 4412 airfoil geometry. The nominal thickness was 12% but was varied, decreasing slightly in thickness moving away from the centerline. Figure 1 shows a top schematic view of the circular wing. The wing geometry was created in CATIA, and later transformed into Initial Graphics Exchange Specification (IGES) file format. The neutral data format is imported to Star Design for preliminary setup. A computational grid of C-H type was generated from this boundary region. Sufficient spacing of 15 diameters between the wing model and the boundary region was selected. The boundary region was defined as free-stream control volume.

3. Meshing

The mesh was decided to generate using polyhedral cells along with prism layer mesh. Table 1 summarizes the overall mesh properties used to represent the model. As shown in Fig. 2., a very fine and high intensity mesh was created on the circular wing in order to preserve and to capture the details of the curvature of the surface as well as the intersection between wing and the domain.

IOP Conf. Series: Materials Science and Engineering **488** (2019) 012005 doi:10.1088/1757-899X/488/1/012005

Table 1: Mesh properties of scaled half-model of the circular wing.

Mesh property	**Parameter**
Mesh type	Polyhedral mesher
Number of prism layers	15
Prism layer stretching	1.3%
Prism layer thickness	15 mm
Cell size (domain)	Max: 0.3 mm, Min: 0.075 mm
Cell size (Circular wing)	Max: 5 mm, Min: 1 mm
Growth rate	1.3%
Total number of cells	1145160

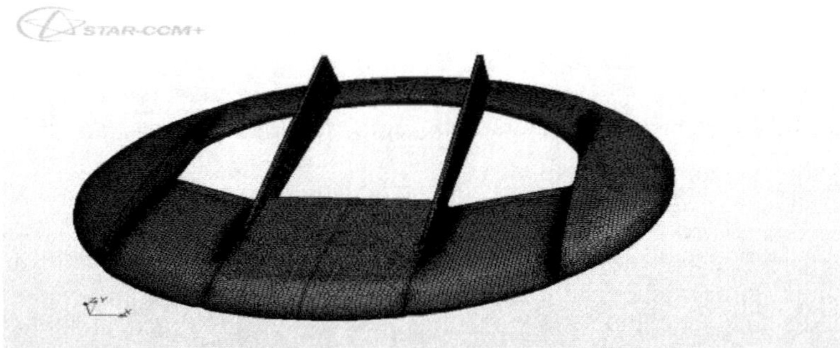

Figure 2. Surface mesh

4. Result and discussion

4.1. Flow visualization

Flow visualization of the circular wing at different angles of attack is presented in Figures 3 to 6 for Reynolds number of $5.1x10^5$. Fig. 3. reveals laminar attached flow with no separation bubble located on the model. There is a high pressure region located on the trailing edge of the elevator as well as separation region near the cockpit and with evident of the existing of detached flow with even more predominant pressure over the cockpit. The flow is completely laminar flow, this is could be the reason for the neutrally stability found in the pitching moment plots.

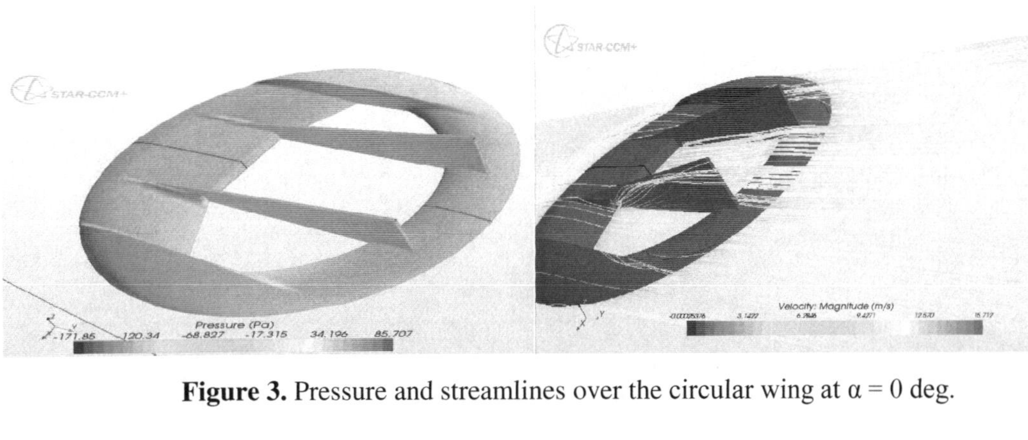

Figure 3. Pressure and streamlines over the circular wing at α = 0 deg.

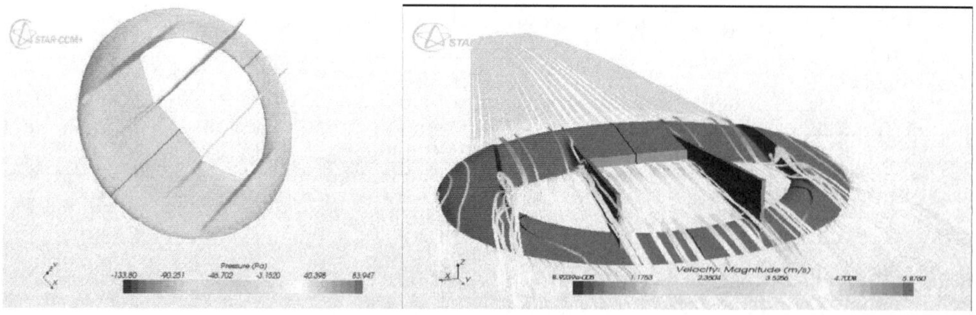

Figure 4. Pressure and streamlines over the circular wing = 10 deg

As the angle of attack is increased to 10 1nd 15 degrees, there is a noticeable change in the structure of the flow with the addition of the separation bubble on the leading edge of the wing. Figure 5 shows this trend but still yields the same pressure on the trailing edges of the control surfaces as well as separation on the top of the cockpit. As α is increased to 15 degrees, several different flow characteristics are observed. Figure 5 shows complete formation of the separation bubble span wise on the airfoil. Existence of a vortex can be seen more prominently at 20 degrees α as shown in Figure 6.

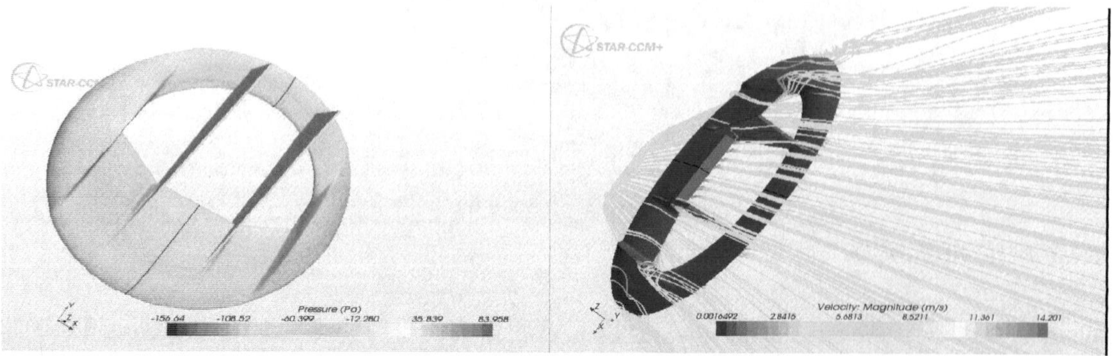

Figure 5. Pressure and streamlines over the circular wing at α = 15 deg.

A complete stall of the aircraft happened at an angle of attack of 27 degrees. Flow on the upper surface has completely changed direction and is moving towards the front of the aircraft.

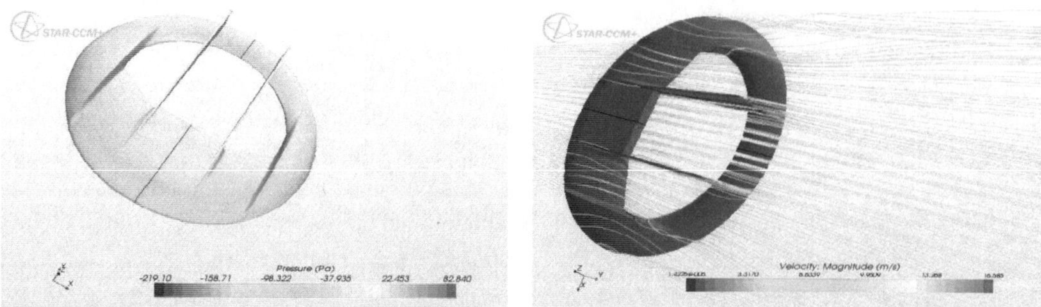

Figure 6. Pressure and streamlines over the circular wing at α = 20 deg

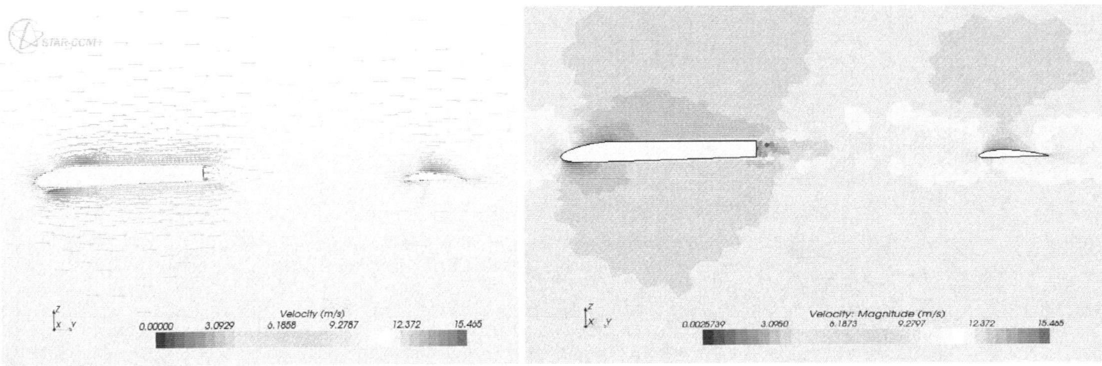

Figure 7. Velocity at y= 0.36m, α =0.

Reynolds number was set as turbulent, which is Re=5.1x10^5 with inflow velocity was set as 12 m/s. The results were performed for angle of attack ranged from 0 degree to 27 degree in order to reach stall angle. However, for this case, simulation was done for two assumptions were considered they are laminar and turbulent since Reynolds number presented transition Reynolds. It can be seen from figures 7 and 10 that the separation region is increased with increasing angle of attack. The separation regions are located at trailing and leading edge of wing.

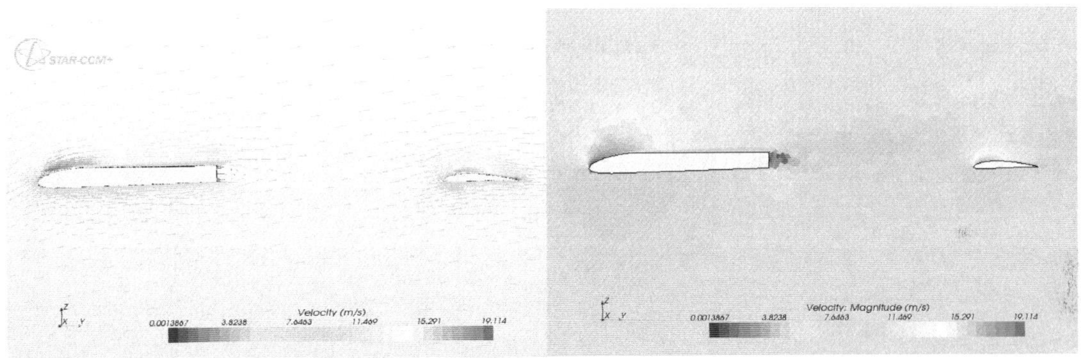

Figure 8. Velocity at y= 0.36m, α =25.

4.2. Aerodynamics coefficients

Figures 9, 10 and 11 represent the aerodynamics force coefficients plot from numerical result for the wing with varies angle of attacks and viscous regime models. As seen from Figure 11, the lift curve slope is larger and has a good stall characteristic, which is at 27 degrees AOA. It also shows that as the lift coefficient increases, the drag coefficient also increases. In term of stability, the wing model can be classified as stable aircraft. Since the experimental testing that carried out by Bryan David (11) was difference Geometry, wing sections and speed but here we can compare the trend for the aerodynamics force coefficient curve distribution. Here, we will only compare the trend for Re=5.1x10^5 as the prototype that built by Recktenwald was tested at the same Reynolds number as this study. Hence, there is only dynamic similarity regardless of Geometric similarity. As shown in Figures 7 and 8, the trends for lift, drag and drag polar curve are almost identical. It was noticed that the wing model that used in this project has better characteristic for stall AOA. In term of stability, both prototype and the wing model share the same characteristic, which is they are statically stable aircrafts.

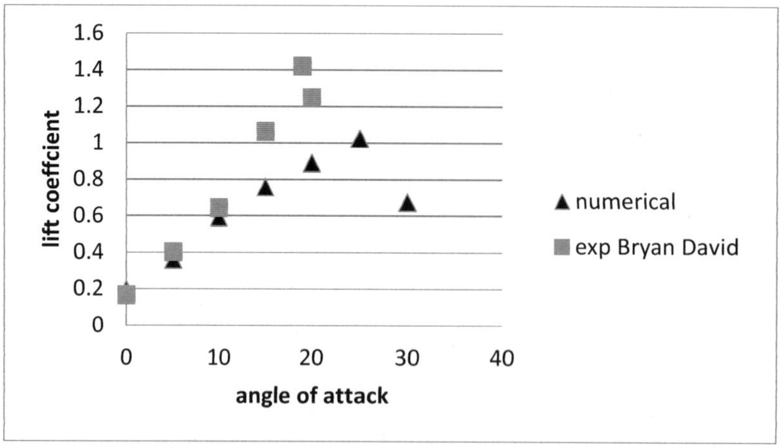

Figure 9. Lift vs angle of attack.

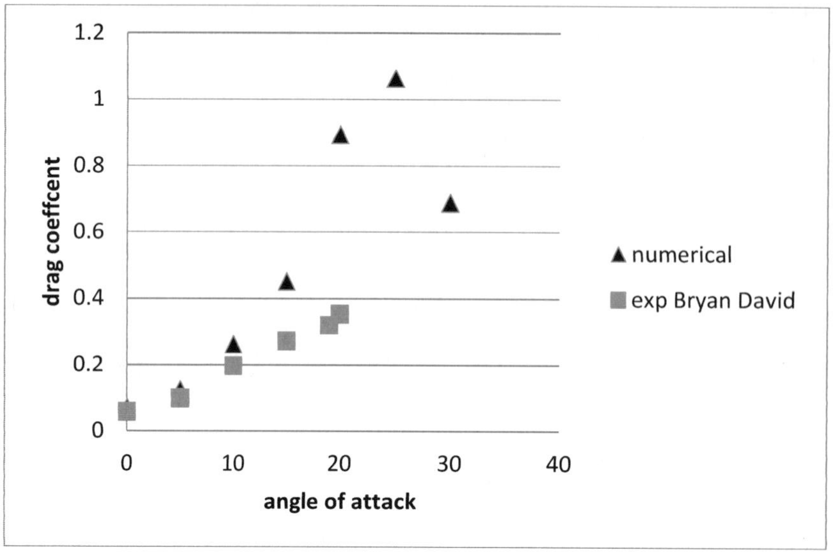

Figure 10. Drag vs angle of attack.

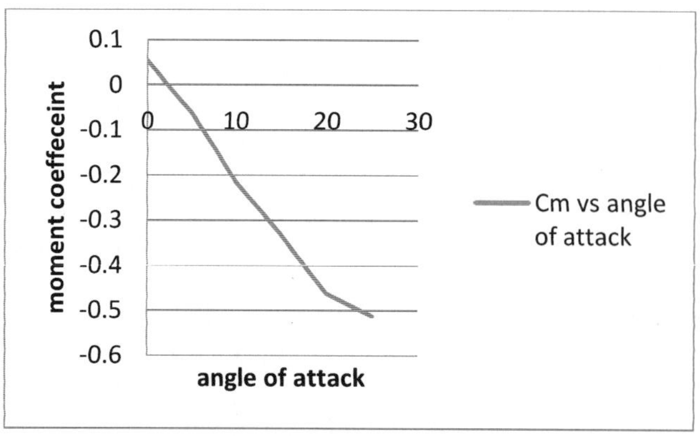

Figure 11. Moment coefficient vs angle of attack

IOP Conf. Series: Materials Science and Engineering **488** (2019) 012005 doi:10.1088/1757-899X/488/1/012005

5. Conclusion

This visualization results shown that the flow was already turbulent in nature. Additional flow structures observed on the circular wing cockpit, control surface trailing edge. The separation vortices increased with angle of attack. The experimental result has shown an agreement with a numerical result as it is clear in figures 3 to 11.

References

[1] Anderson JD. 1999 Aircraft performance and design. *Vol. 1. WCB/McGraw-Hill Boston*, MA

[2] Anderson Jr JD. 2010 *Fundamentals of aerodynamics. Tata McGraw-Hill Education.*

[3] Von Kaenel R, Rizzi A, Oppelstrup J, Goetzendorf-Grabowski T, Ghoreyshi M, Cavagna. 2008 simulating stability & control with CFD/CSM in aircraft conceptual design. *In: 26th International Congress of the Aeronautical Sciences,* ICAS, pp. 14–9

[4] Wilson T. 2008 *Americas Nuclear Flying Saucer*. Pop Mech.

[5] Raymer DP, Zhang M, Rizzi AW, Raymer EN. 2017 Sparky Flapjack: Aircraft Design Inspirations from the Vought 173. *In: 55th AIAA Aerospace Sciences Meeting*, pp. 1184.

[6] Suliman MM, Omar AA, Syed Mohamed Ali J. 2013 Flow visualization over a circular planform wing. *In: 2nd International Conference on Mechanical, Automotive and Aerospace Engineering (ICMAAE' 13), 2-4 July 2013, Kuala Lumpur, Malaysia.*

[7] Stilley GD. 1972 Aerodynamic analysis of the self sustained flair. AD-740117, *Nav Ammunit Depot, Crane, Ind.*

[8] Recktenwald B. 2008 Aerodynamic Testing of a Circular Planform Concept Aircraft.

IOP Conf. Series: Materials Science and Engineering **488** (2019) 012006 doi:10.1088/1757-899X/488/1/012006

Modelling the performance and emission prediction of RB211 aero-gas turbine engine fuelled by Jatropha-based biofuel

Muhammad Hanafi Azami[1], Zahid Noorazman[2], Mark Savill[3], Yi-Guang Li [4] , and Mohd Razi Hilmi[5]

[1,2]Department of Mechanical Engineering, Kulliyyah of Engineering, International Islamic University Malaysia
[3,4]Centre for Propulsion Engineering, Cranfield University, United Kingdom
[5]Department of Optometry and Visual Sciences, Kulliyyah of Allied Health Sciences, International Islamic University Malaysia

*Corresponding author e-mail: hanafiazami@iium.edu.my

Abstract. Fossil fuel is one of the world vital energy resources. The development of transportation technologies increases the demand for petroleum derivative globally. Fossil fuel consumption produces emissions, which potentially harm the environment and human health. Many mitigations have been implemented to address the two main crises; the energy scarcity and environmental calamity. This paper will discuss on one of the potential solutions by analyzing the performance and emission prediction of aero-gas turbine engine fuelled by Jatropha-based biofuel. Performance analysis was made based on the thrust and specific fuel consumptions at different blended ratio percentages for various flight conditions. The three-shaft high-bypass-ratio engine model, which is identical to the Rolls Royce RB211-524 was used to model in an in-house Cranfield's University software, PYTHIA. PYTHIA is integrated with the TURBOMATCH performance evaluation programme by iterating the mass and energy balance for each engine component. The analysis is then continued to predict Nitrogen Oxides emission index (EINOx) at every flight conditions using an in-house Cranfield's University computer tool, HEPHAESTUS. HEPHAESTUS is an emission prediction software by using Zel'Dovich equations (for NOx) and models the emission by implementing a partially-stirred reactor (PSR) model and perfectly stirred reactor (PSRS) models at different zones in the combustor. Validation showed that HEPHAESTUS is able to capture a reasonable prediction as compared to the International Civil Aviation Organization (ICAO) databank. The performance the biofuel has shown an improvement in engine performance at higher percentage blended ratio but also increase the nitrous oxide indices emission slightly.

1. Introduction

Fossil fuels are highly demanded globally. Burning the fuel is one of the oldest methods for generating energy and power all around the world. Fossil fuels (oil, natural gas, and coal) contribute to approximately 80% of total world energy supply [1]. The World Energy Outlook (WEO) 2007 claims that in 2030, it is expected that the energy demand will grow up to 84% as the fossil derivative will remain to be the major source. By 2030 fossil fuel is predicted to amplify the quantity up to 61 million barrels per day from 36 million barrels per day in 2006 [2]. The amount of all three types of fossil derivative escalate as the time increases. Besides, the demands on oil are always the highest, while the demands on coal are greater than gas. Consequently, the consumption of fossil fuels has grown over the last 40 years and is expected to follow the same trend in the future.

The world fossil derivative resources are estimated to reach the peak between 2013 and 2020 [3]. Through the formalisation of the depletion of fossil fuel rate curves, it is a hesitate to believe that the symmetrical bell curve is the last resort in predicting the completion of the fossil derivative lifespan [4]. The example of the country that has found the fossil fuel production peak has already been reached is Turkey. Turkey has indicated that their fossil derivative will diminish in 2038 [5]. This

IOP Conf. Series: Materials Science and Engineering **488** (2019) 012006 doi:10.1088/1757-899X/488/1/012006

shows that the fuel reserve is depleting rapidly as the global demand hike up. Moreover, fossil fuel utilization also produces emissions that can contribute harm towards the health at the human population level as well as the environmental concerns. The Kingdom of Saudi Arabia reported its emissions of CO_2, CH_4, and N_2O from road transportation based from fossil fuel in 2000 are approximately 56 million tons, 10.42 thousand tons, and 0.47 thousand tons, respectively [6]. In 2010, the estimated emissions of CO_2, CH_4, and N_2O are approximately 98.12 million tons, 18.19 thousand tons, and 0.83 thousand tons, respectively. In 2020, the forecasted emissions of CO_2 are expected to be 163.33 million tons, CH_4 are expected to be 32.72 thousand tons and N_2O are expected to be 1.39 thousand tons [7].

Biofuel can be derived from biomass for instance vegetable oils such as palm oil, rapeseed oil and others [8]. The raw materials for the preparation of biodiesel can be classified into three types: (i) Vegetable oils; (ii) Waste frying oils and (iii) Animal fats. The main superiority of biofuel is that an existing internal combustion engine can operate directly by utilizing it without major modifications. Jatropha biofuel is one of the alternatives solutions towards rapid depletion of fossil derivatives due to Jatropha can be reproduced in a short period of time. Therefore, it can potentially reduce the global dependence on fossil derivatives. Meanwhile, as in environmental concern, the production of Jatropha biodiesel releases less greenhouse gas (GHG) emissions as compared to conventional fossil derivatives. Alongside, the biofuel can be reproduced over a short period, the prediction has shown that the use of Jatropha biofuel can abbreviate the greenhouse gas emissions up to 23% from the conventional fossil derivative [9]. Furthermore, Jatropha Curcas plantation is very economical as it natures of scarcity crops. With a little quantity of moisture, it can grow on rough land and can keep producing up to 40 years [10]. Jatropha crops also need no fertilizers, crop rotation. no insecticides also required. has the lowest cost compared to other biomass feedstocks, and inedible plant. This shows that Jatropha-based biofuel can potentially be a source of renewable energy and as an alternative to conventional fossil derivative while simultaneously reduce harm on the environment. For this reason, an engine modelled fuelled by Jatropha-based biofuel is studied. The research is directed towards the modelling of the performance and emission in a conventional aero-gas turbine engine (Rolls Royce RB211-524).

2. Methodology

The Jatropha Curcas based biofuel will be analysed as pure Jatropha biofuel, pure kerosene and also blended Jatropha-Kerosene at 20%, 40%, 60% and 80% percentage blended ratio. Continental Airlines, Air New Zealand and Interjet are among the airlines' companies that used Jatropha as the alternative fuel. A model of a 3-shaft high-bypass-ratio engine RB211-524 was used. For simulation and computational analysis, an in-house Cranfield's University software named PYTHIA program is used. Aero-gas turbine engine model and gas path analysis in PYTHIA apply a modified Newton-Raphson convergence technique. In addition, it can fill in as indicative apparatus for disintegration evaluation for off-design conditions. TURBOMATCH is a coded program that is incorporated with PYTHIA. PYTHIA has an interactive interface which requires user input then it calls for TURBOMATCH for the iteration and matching each engine components through mass and energy balance equations.

The advantage of the new version of PYTHIA program is that it is capable to vary the running fuel type as well as blending two different type of fuel with the variation of blended ratio percentages. Therefore, this work could be therefore served as an expansion of [11] work using an earlier version of PYTHIA which could only provide strict comparisons for different pure fuels for single design conditions only. Rolls Royce RB211-524 aero-gas turbine engine model is chosen in the PYTHIA's library. At a design point, kerosene fuel was selected. Each component of the engine model is described in terms of a 'brick' which has its own functionality. The details of the 'bricks' can be found in [12]. The sequential diagram of PYTHIA process can be found in [12]. It begins with the user defining inputs as previously mentioned in PYTHIA. TURBOMATCH is called for the iterations in mass equation (1) and energy equation (2) balance relation. Equations (1) and (2) should be satisfied between successive components.

$$\frac{W_n\sqrt{T_n}}{P_n} = \frac{W_{n+1}\sqrt{T_{n+1}}}{P_{n+1}} \quad (1)$$

$$Turbine\ Work\ (TW) = Compressor\ Work\ (CW) \quad (2)$$

The empirical method is often used to predict Nitrogen Oxides, NOx. The parameter of an aero gas turbine NOx emission can be determined through its emission index (EINOx). Some of the empirical correlations used in the literature in [13]. For simulation and computational analysis, this research uses in-house computer software which is called HEPHAESTUS program. It begins by inserting a few input data gained from PYTHIA software in combustor brick. The parameters that are obtained from performance PYTHIA simulation part and necessary for the simulation are ambient (flight) altitude, ambient temperature, air total temperature, air total pressure, total air mass flow rate and fuel mass flow rate. While others parameter that has been fixed based on the specifications of the Rolls Royce RB211-524 aero gas turbine as in [11].The forecasted emissions simulation is for pure kerosene, pure Jatropha and blended Jatropha and kerosene. Even though HEPHAESTUS is capable to forecast many biofuels emissions easily, only EINOx are focused for this paper.

3. Results and Discussion

The Rolls Royce RB211-524 aero gas turbine model designed in PYTHIA was validated by comparing with [7] experimental work earlier. The test was successfully conducted on Air New Zealand airline by running 50% Jatropha and 50% Jet-A fuel on Boeing 747-400 [14]. Figure 1 shows the performance in term of thrust at different flight speeds for five different blended ratio percentages of Jatropha biofuel and kerosene. Generally, all blended ratio of KE+BJ biofuel recorded an increase in thrust. The result indicates that blended Jatropha biofuel can improve the performance of aero gas turbine in thrust up to 0.47%(at 100% blended ratio and 0.4 Mach number). Meanwhile, Figure 2 shows the performance in term of Specific Fuel Consumption(SFC) at different blended percentage ratio. The results recorded a reduction of SFC at higher blended percentage ratio. It indicates that blended Jatropha-based biofuel can improve the performance of aero gas turbine in SFC reduction up to 2.69%(at 100% blended ratio and 0.8 Mach number).

Nitrous oxides formation has a complex chemical reaction. Some of empirical methods and correlation are normally used for predicting NOx in term of NOx emission index (EINOx). There are 8 correlations can be compared, which are, LIPFERT, AECMA, GASTURB, LEFABRE, BLAZOWSKI, NASA, and NPSS. The correlations can be found in [13]. Figure 3 predicts the variation of NOx emissions using different correlations at different blended ratio percentages. The HEPHAESTUS program recorded an increment in EINOx as the blended ratio moves towards pure biofuel (highest blended ratio). It is shown that HEPHAESTUS prediction falls within the range of all 8 correlations for prediction EINOx method, thus, HEPHAESTUS prediction is acceptable.

IOP Conf. Series: Materials Science and Engineering **488** (2019) 012006 doi:10.1088/1757-899X/488/1/012006

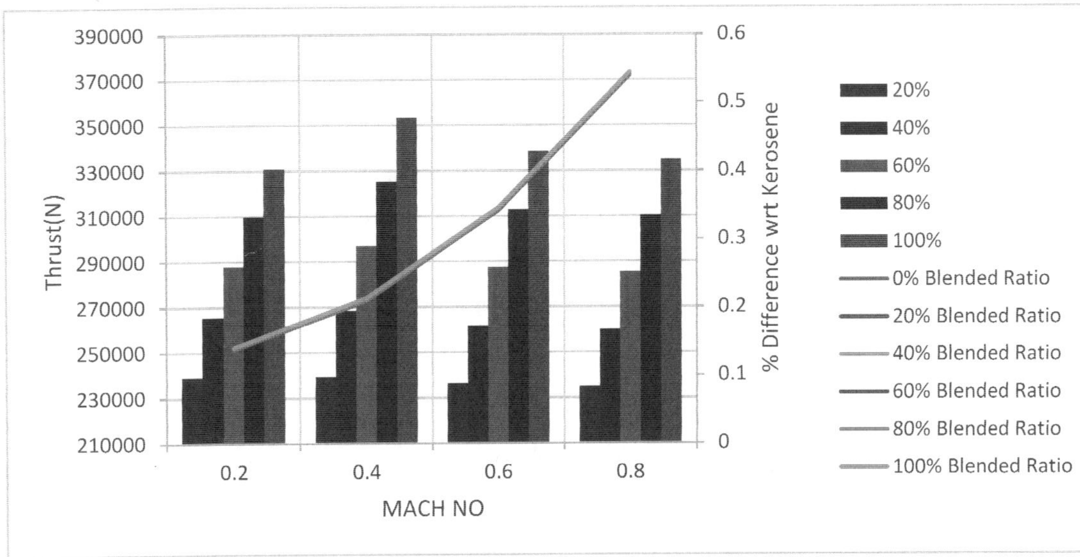

Figure 1. Thrust VS Mach Number

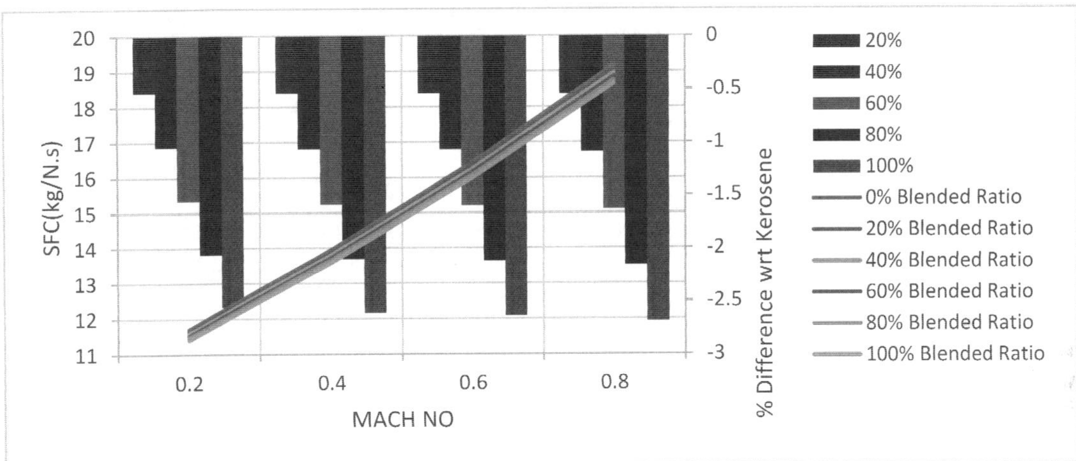

Figure 2. Specific Fuel Consumption VS Mach Number at take-off condition.

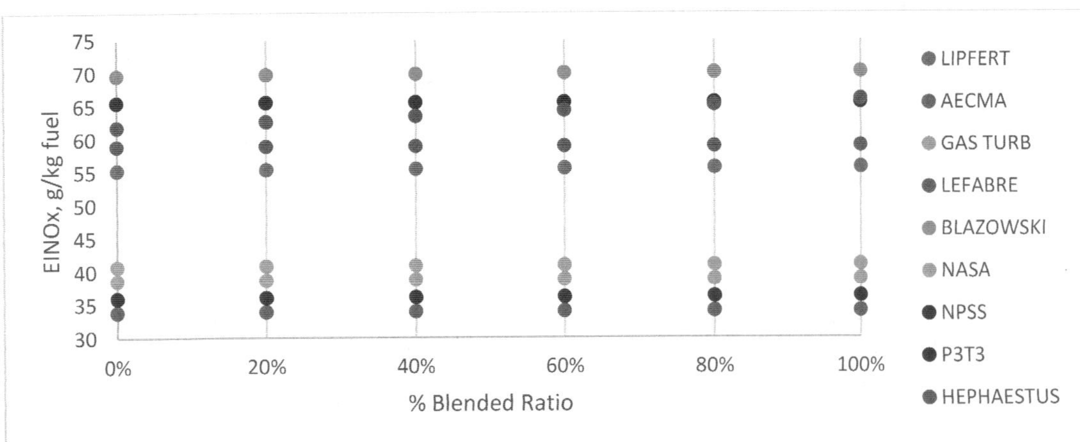

Figure 3. EINOx vs Blended Ratio at Take-off condition

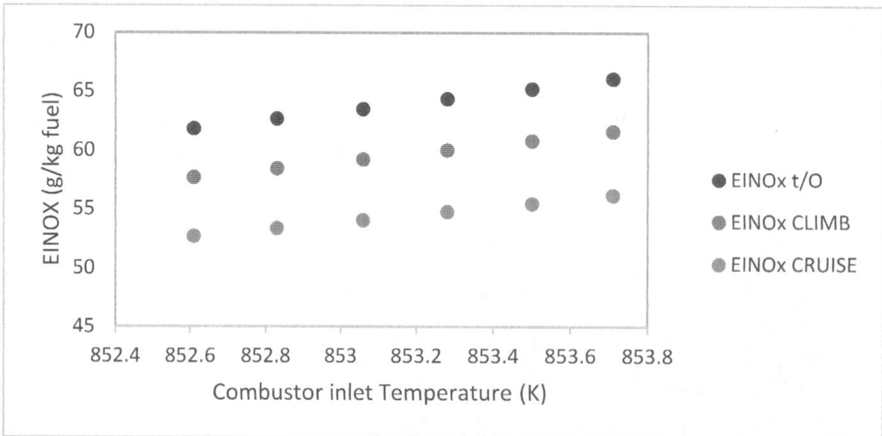

Figure 4. EINOX (g/kg fuel) vs Combustor Inlet Temperature (K)

Figure 4 shows emissions prediction using HEPHAESTUS at various flight conditions. Generally, all flight condition of KE+BJ biofuel has the same pattern which recorded an increment in EINOx due to the increase in combustor inlet temperature. Take-off flight condition resulted in a high amount of NOx due to throttling and burning more fuel. High temperature resulted in more NOx formation. NOx formation is highly dependent on the temperature.

4. Conclusion

In conclusion, Jatropha-based biofuel has been proved to be one of the most potential crops in biofuel industries. The simulation using PYTHIA software has shown that Jatropha biofuel can improve aero gas turbine performance in term of thrust and specific fuel consumption reduction up to 0.47% and 2.69% respectively. Furthermore, modelling of the Nitrogen Oxides emission index (EINOx) prediction of aero-gas turbine utilizing Jatropha biofuel successfully conducted using HEPHAESTUS codes.

Acknowledgement
This work was funded by RIGS Grant (RIGS17-106-0681) of International Islamic University Malaysia.

References

[1] Goldemberg J 2007 Ethanol for a Sustainable Energy Future **315**, February, pp. 808–810.

[2] Ramsay W. C. 2007 World Energy Outlook 2007: China and India Insights.

[3] M. G. Salameh Can renewable and unconventional energy sources bridge the global energy gap in the 21st century? **75**, pp. 33–42, 2003.

[4] Seifritz W. 2003 An endogenous technological learning formulation for fossil fuel resources **28**, pp. 1293–1298.

[5] Volkan S. 2007 ARIMA forecasting of primary energy demand by fuel in Turkey **35**, pp. 1701–1708.

[6] Zatari T.M 2011 Second national communication: Kingdom of Saudi Arabia.

[7] Rahman S. M., Khondaker A. N., Hasan A., and Reza I. 2016 Greenhouse gas emissions from road transportation in Saudi Arabia - a challenging frontier *Renew. Sustain. Energy Rev.*, **69**(6), pp. 812–821.

[8] Hsieh W., Chen R., Wu T., and Lin T. 2002 Engine performance and pollutant emission of an SI engine using ethanol – gasoline blended fuels **36**, pp. 403–410.

[9] Mofijur M. *et al.* 2012 Prospects of biodiesel from Jatropha in Malaysia *Renew. Sustain. Energy Rev.*, **16**, no. 7, pp. 5007–5020.

[10] Dubey P. and Gupta R. 2018 Influences of dual bio-fuel (Jatropha biodiesel and turpentine oil) on single cylinder variable compression ratio diesel engine *Renew. Energy*, **115**, pp. 1294–1302.

[11] N. Musfirah 2015 Effects of biofuels properties on aircraft engine performance Cranfield University Thesis .

[12] Muhammad Hanafi Azami, Mark Savill 2017 Comparative study of alternative biofuels on aircraft engine performance. *Proceedings of the Institution of Mechanical Engineers, Part G: Journal of Aerospace Engineering,* **231** (**8**). , 0 pp. 1509-1521. ISSN 0954-4100

[13] Chandrasekaran, N. and Guha, A. 2012 Study of Prediction Methods for NOx Emission from Turbofan Engines *Journal of Propulsion and Power*, **28(1)**, pp. 170–180. doi: 10.2514/1.B34245..

[14] Rahmes, T. F., Kinder, J. D., Henry, T. M., Crenfedlt, G., LeDuc, G. F., Zombanakis, G. P., Abe, Y., Lambert, D. M., Lewis, C., Juenger, J. A., Andac, M. G., Reilly, K. R., Holmgren, J. R., McCall, M. J. and Bozzano, A. G. 2009 Sustainable Bio-Derived Synthetic Paraffinic Kerosene (Bio- SPK) Jet Fuel Flights and Engine Tests Program Results in *9th AIAA Aviation Technology, Integration, and Operations Conference (ATIO)*. Hilton Head, South Carolina: American Institute of Aeronautics and Astronautics, pp. 1–19. doi: doi:10.2514/6.2009-7002.

IOP Conf. Series: Materials Science and Engineering **488** (2019) 012007 doi:10.1088/1757-899X/488/1/012007

Modelling the performance of aero-gas turbine engine using algae-based biofuel with emission prediction

Muhammad Hanafi Azami[1], Muhamad Zaki[2], Mark Savill[3], and Yi-Guang Li[4]

[1,2]Department of Mechanical Engineering,
Kulliyyah of Engineering, International Islamic University Malaysia
[3,4]Centre for Propulsion Engineering,
Cranfield University, United Kingdom

Corresponding author e-mail: hanafiazami@iium.edu.my

Abstract. The world oil consumption is at the peak where the fuel price is insubstantial and can increase dramatically due to economic, social, and political factors and unprecedented stability. Since fuel resources are scarce, it is an urgent need to find alternative fuel. Biofuel is one of the favorable choices in the market. Algae-based biofuel is the fourth generation of biofuel where it does not compete with the food production and it has myriad of advantages. These abundant algae are easy to cultivate and researchers found that algae-based biofuel is capable of reducing engine emission. This paper modelled the RB211 aero-gas turbine engine by utilizing algae-based biofuel with various blended percentage ratios at different flight conditions. Cranfield's University in-house software, PYTHIA, and HEPHAESTUS are used to model the engine performance and emission prediction respectively. PYTHIA programme uses a modified Newton-Raphson convergence technique in the zero-dimensional steady-state model for both design and off-design conditions. Meanwhile, HEPHAESTUS software uses the Zeldovich equations (for NOx) and models the emission by implementing a partially-stirred reactor (PSR) model and perfectly stirred reactor (PSRS) models at different zones in the combustor. Results have shown that thrust force produced is increasing at higher blended percentage ratio of algae biofuel. Through emission analysis prediction, generally, the nitrous oxide emission formation is lower at a higher altitude during the cruising. Results also predicted that higher percentage blended ratio of algae biofuel also reduces the emission formation.

1. Introduction

Nowadays, the technology advancement in aircraft engine have experienced a dramatic increase. Engineers are studying on how to improve the performance of the engine and make it better every day. The most interesting topic that engineers and also citizens care about is fuel efficiency and emission caused by transportation. In about 10 years to the future, we will see about 15-20 percent improvement in fuel efficiency. A study by Daggett [1] on fuels in commercial aircraft said that "air travel growth is predicted to continue at 5 percent per year". This means that the future fuel efficiency will be outpaced by increasing amount of fuel needed in aircraft industry. Thus, we can predict the oil peaking will occur one day and we have less energy resources available to use.

Aircraft emission also responsible in rising concentration of greenhouse gases (GHGs) in the atmosphere. It is reported by Anger [2] that aviation industry produces carbon dioxide emission that cover about 3% of global emission in 2004. Literally the number is small, however the European environment Agency [3] reported that carbon dioxide emission from aviation in European countries showed a massive growth from 1990 to 2004, about 85%. These two data concluded that number of emission every year keep increasing and should be a future concern. Besides, emission from aircraft engine also include nitrogen oxides, sulphur dioxide, soot and carbon monoxide. Anger [2] added that, emissions from aircraft engine that are less well understood is the formation of cirrus cloud and contrails. Cirrus cloud are in form of high level clouds and have greenhouse effect and solar albedo

IOP Conf. Series: Materials Science and Engineering **488** (2019) 012007 doi:10.1088/1757-899X/488/1/012007

effect [4]. Contrails are line shaped clouds form that are visible from the movement of aircraft in air. The impact of contrails is quite similar to impact by carbon dioxide emission, reported by Penner[5]. It is understood, that current aircraft condition is in need of sustainable energy for future use that are renewable and produce low emission.

In these generation, the research on biofuel as alternative fuel has been an interesting discussion among researchers. Biofuels or plant derived fuel offer a wide variety of selection such as soybean oils, rice bran oil, palm oils, corn, algae and many more. This allows us to do various kind of experiment to test the effectiveness of this biofuel. The most promising biofuel as an alternative fuel is algae. It grows abundantly and does not require agricultural land. Algae has huge supply and could be the future fuel for aircraft operation as research also find that this biofuel could reduce emissions in aircraft operation. The most common type of algae is green algae (chlorophyta), red algae (rhodophyta) and brown kelps (Phaeophyta). Based on a research by He [6], green microalgae is more suitable for biofuel production due to high lipid content than other type of microalgae. It has about 25% wt lipid content and high lipid content could improve the efficiency of biomass processing. It is also important to note that green algae are mostly freshwater species. Brown microalgae has high carbohydrate content and high ash content. Thus, brown microalgae is not good for biodiesel production and more suitable for fermentable sugar production.

Biofuels that are developed for aircraft applications are called "biojet" fuels. There are several challenges in using this biojet such as its tendency to freeze at normal operating flight temperatures until recent development improved this. Other major challenge is this biofuel has poor high temperature thermal stability characteristics. The biofuel need to blend with major percent of aircraft fuels (jet-A).

This research is focusing on modelling gas turbine engine in various conditions. This include varying fuel mixture and fuel type. Thus, software chosen to simulate the result must be able satisfy this requirements and user friendly. PHYTIA software is selected as it can be used to stimulate engine performance and considered to have user friendly interface and been tested for many years in validation of numerous gas turbines [7]. The data obtained from this software is thrust, specific fuel consumption and others. In emission modelling, HEPHAESTUS software will be used as it uses the same data from PHYTIA and can predict emissions from biofuel with ease [8]. The main data required from PHYTIA software is emission index EINOX.

2. Methodology

The PHYTIA program was selected for computational analysis. Igie [9] stated that this program can design and measure numerous gas turbine engines for both design and off-design points using a modified Newton–Raphson convergence technique in the zero-dimensional steady-state model. PHYTIA is integrated with TURBOMATCH code to solve iterations in mass equation and energy balance for engine component selected [8].

$$\frac{W_n\sqrt{T_n}}{P_n} = \frac{W_{n+1}\sqrt{T_{n+1}}}{P_{n+1}} \tag{1}$$

$$Turbine\ Work\ (TW) = Compressor\ Work\ (CW) \tag{2}$$

Then, the data obtained are imported to the excel spreadsheet for data analysis as shows in Figure 1. PHYTIA is considered to have user friendly interface and been tested for many years in validation of numerous gas turbines [10]. The latest update on PHYTIA enable it to change the fuel type and diversify the blended mixing ratio percentage while using same engine selection for the kerosene case [11]. This enables deeper research on fuel usage for engines in different operating condition.

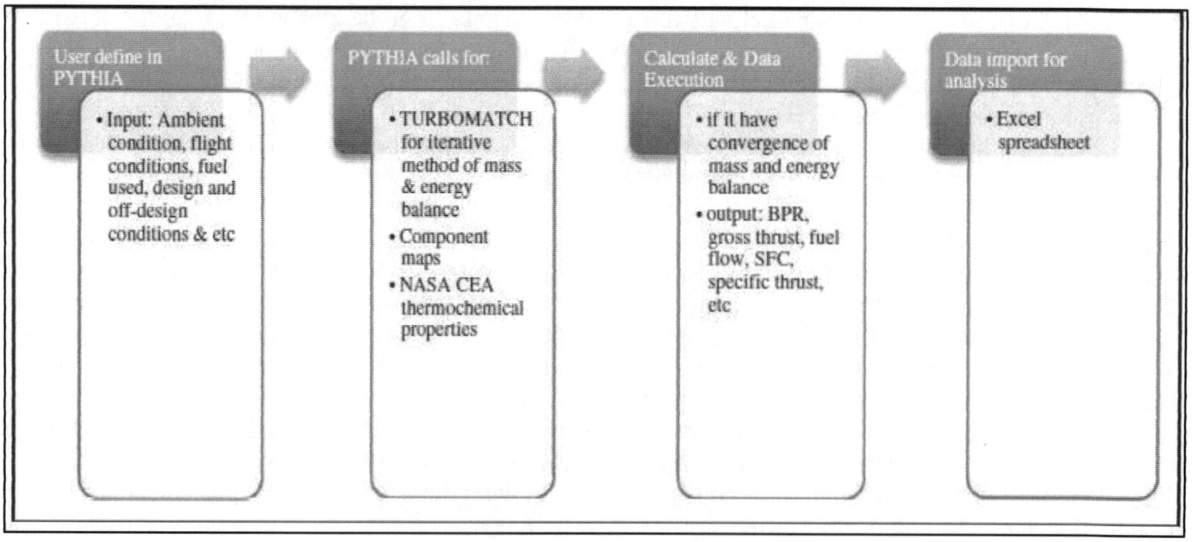

Figure 1. PHYTIA process flow chart. [8]

Software HEPHAESTUS are used in emission analysis for this research as it can predict carbon monoxide and nitrous oxides emissions at multiple flight conditions. This software required data that were obtained from PHYTIA. The parameter required for the software input are fuel total temperature, ambient conditions, air total pressure and temperature at the combustor inlet, fuel and total air mass flow rate. Emission prediction technique used to validate the data obtained by HEPHAESTUS is called correlation-based models. It is among the simplest method to predict the emission, however it require large number of data from the engine/combustor. Primary variables such as pressure and temperature are obtained from PHYTIA.

3. Result and discussion

Figure 2 shows that the thrust increase along with the increase of blended ratio percentage at the altitude of 2000m. At higher Mach numbers of 0.6 and 0.8, the difference in thrust has increases significantly as blended percentage ratio increases. In terms of specific fuel consumption, SFC which is used to describe the fuel efficiency of an engine design to thrust output. SFC for gas turbine is the mass of fuel needed to provide the net thrust for a given period. With the same flight conditions and engine configurations the SFC can be identified. Figure 3 shows the SFC trends and the percentage difference of SFC with Mach number. Unfortunately, the SFC has increase at higher percentage blended ratios. This shows that the algae-based biofuel consumed more fuel compared to kerosene fuel.

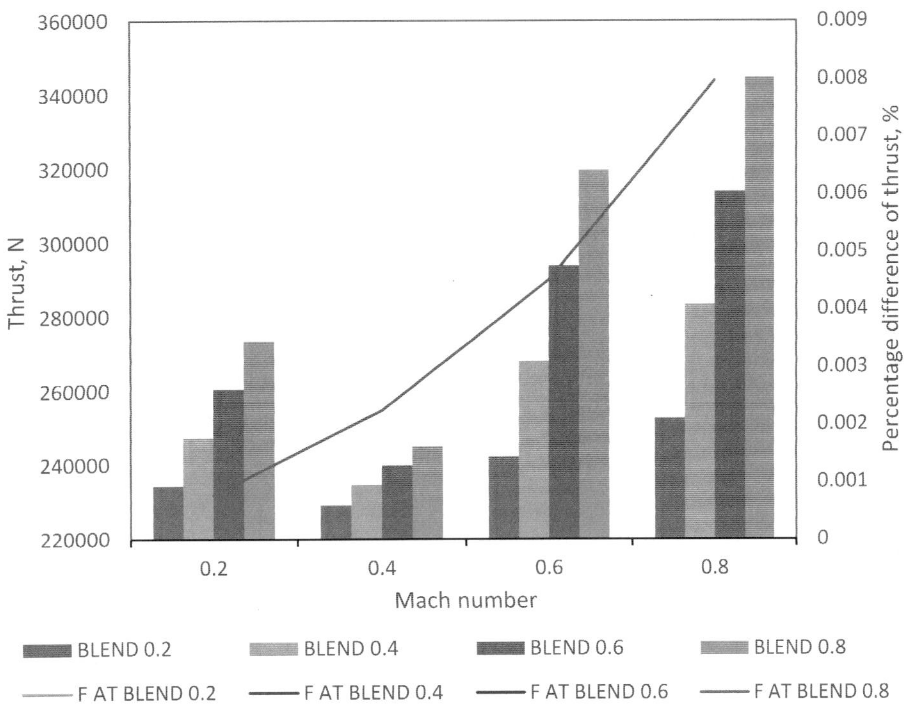

Figure 2. Thrust vs Mach number with respect to blended ratio at altitude 2000

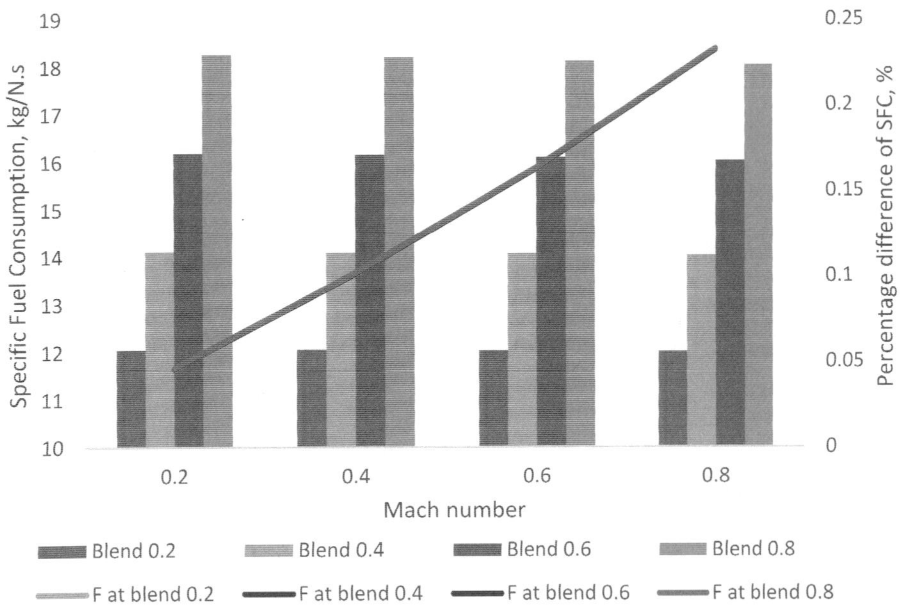

Figure 3. Specific fuel consumption (SFC) Vs Mach number with respect to blended ratio at altitude 2000

IOP Conf. Series: Materials Science and Engineering **488** (2019) 012007 doi:10.1088/1757-899X/488/1/012007

Prediction of engine NOx emission is complex and it is measure in term of NOx emission index (EINOx). Emission model prediction from HEPHAESTUS are compared with published correlation based models such as Lipfert and Lefebvre correlations. The results are tabulated in Table 1. According to Chandrasekaran [12], Lipfert correlation over predict EINOx value for all gas turbine engine analyzed. This means that the value is a little bit diverged from other data but still in the acceptable range. In addition, the percentage difference between HEPHAESTUS value and the two correlation stated is under 10%. Thus, it can be assumed that the EINOx value from HEPHAESTUS is accepted and is the main data.

Table 1. EINOx analysis at Mach 0.4 at altitude 2000

Blended ratio (%)	Lipfert and % difference with HEPHAESTUS	Lefebvre and % difference with HEPHAESTUS	Hephaestus (g/kg fuel)
0	52.79 (7.68%)	56.88 (0.54%)	57.19
100	52.81 (6.78%)	57.47 (1.47%)	56.65

It shows that the HEPHAESTUS has an acceptable trend, and the capability of HEPHAESTUS has been extended to predict the nitrous oxide emission at various altitudes. Figure 4 shown that the EINOx of algae-based biofuels has much lower EINOx and the trend is decreasing at higher altitude. It is believed that EINOx trend is reduces in algae-based biofuel due to the lower flame temperature of the fuel. EINOx is highly sensitive to temperature as well as pressure. In PYTHIA, we observed that the combustion chamber temperature is much lower using algae-based biofuel.

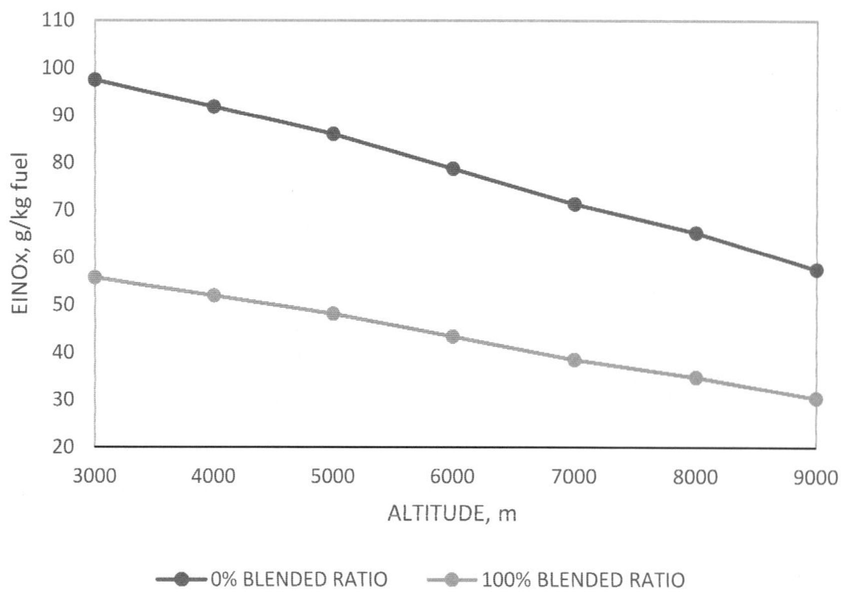

Figure 4. EINOx at cruise speed vs altitude with respect to KE+AG blended ratio

4. Conclusion

The effect of different blended percentage ratio of algae-based biofuel to performance and emission of gas turbine are determined. Results shown that the thrust force is increasing at higher blended percentage ratio. However, the SFC also tends to increase as well. This might be the disadvantage. For the emission analysis prediction, shown in HEPHAESTUS, the nitrogen oxide emission index can be reduced using algae-based biofuel.

Acknowledgement

This work was funded by RIGS Grant (RIGS17-106-0681) of International Islamic University Malaysia.

References

[1] Daggett DL, Hendricks RC, Walther R, and Corporan E 2008 Alternate Fuels for Use in Commercial Aircraft.

[2] Anger A. 2010 Including aviation in the European emissions trading scheme: Impacts on the industry, CO2 emissions and macroeconomic activity in the EU **16**, no. 2, pp. 100–105.

[3] EEA environmental statement European Environment Agency Report 2007 Kongens Nytorv 6 1050 Copenhagen, Denmark.

[4] Zerefos CS, Eleftheratos K, Balis DS, Zanis P, Tselioudis G., and Meleti C. 2003 Evidence of impact of aviation on cirrus cloud formation *Atmos. Chem. Physics,* **3(5)** no. 1996, pp. 1633–1644.

[5] Penner JE, Lister DH, Griggs DJ, Dokken DJ, and McFarland M 1999 Aviation and The Global Atmosphere: A Special Report of IPCC Working Groups *Intergov. panel Clim. Chang.*, p. 373.

[6] He BB 2012 In situ Transesterification of Microalgal Oil to Produce Algal Biodiesel, Moscow.

[7] Azami MH, Savill M., and Li Y 2017 Comparison of aircraft engine performance and emission analysis using alternative fuels *Int. J. Sustain. Aviat. Vol. 3, No. 1*, **3**, pp. 1–21.

[8] Azami MH and Savill M 2017 Comparative study of alternative biofuels on aircraft engine performance *Proc. Inst. Mech. Eng. Part G J. Aerosp. Eng.*, vol. **231**, no. 8, pp. 1509–1521.

[9] Igie U. and Minervino O. 2014 Impact of Inlet Filter Pressure Loss on Single and Two-Spool Gas Turbine Engines for Different Control Modes *J. Eng. Gas Turbines Power*, vol. **136**, no. 9, p. 091201.

[10] Li YG *et al.* 2012 Improved Multiple Point Nonlinear Genetic Algorithm Based Performance Adaptation Using Least Square Method *J. Eng. Gas Turbines Power*, vol. **134**, no. 3, p. 031701.

[11] Mazlan NM, Savill M, and Kipouros T 2015 Effects of biofuels properties on aircraft engine performance," *Aircr. Eng. Aerosp. Technol.*, vol. **87**, no. 5, pp. 437–442.

[12] Chandrasekaran N and Guha A 2012 Study of Prediction Methods for NOx Emission from Turbofan Engines," *J. Propuls. Power*, vol. **28**, no. 1, pp. 170–180.

IOP Conf. Series: Materials Science and Engineering **488** (2019) 012008 doi:10.1088/1757-899X/488/1/012008

Review of Manufacturing Process for Good Quality of Composite Assessment

II Zulkepli[a]*, H Mokhtar[b], Y Aminanda[c], M S I Shaik Dawood[d] and M S M. Rehan[e]
[1]Department of Mechanical Engineering
Kulliyyah of Engineering, International Islamic University Malaysia
Jalan Gombak, 53100 Kuala Lumpur, Malaysia

E-mail: [a]intan.irieyana@yahoo.com.my,[b]hananm@iium.edu.my, [c]sultan@iium.edu.my, [d]yulfianaminanda@gmail.com, [e]msaifuddin@iium.edu.my.

Abstract. This review paper focuses on the detailed technique and process of fiber fabrication until it becomes composite which is useful for the manufacturing industry. The most famous fabrication methods are the open and closed mould. In general, open mould is a method which exposes the specimen to the atmosphere and the surface condition is not controlled. Close mould, however, is the opposite to open mould, that is in closed conditions without direct contact of the atmosphere and the surface condition is controllable. The manufacturing process techniques are presented comparatively. Discussion on fiber type that is more suitable for specific techniques are also presented in this paper.

1. Introduction

Treating such as heat treating or coating, machining, reshaping, are some of the techniques in manufacturing process. Furthermore methods on the testing, checking and planning are necessary to inspect the quality assurance during or after manufacturing [1]. Many developments appear to fabricate the fiber however it must meet the specific designs or manufacturing challenges [2]. Fiber reinforcement in composite has to go through composite fabrication process. The process normally involved metals or plastics by moulding or shaping mould into polymer group. Polymers commonly also known as plastics and most widely used in matrix materials especially in fiber composites. Polymer is the 'glue' used to hold together the high stiffness, high-strength fiber to become as matrix. It is popular due to its low cost, easy processibility, good chemical resistance, excellent corrosion resistance, low specific gravity and simple fabrication methods [3]. There are two classification of polymer that dominating matrices offered highest potential of composite structure which is thermoplastics and thermosets. Thermoplastics can be easily melted or softened when applied to heat and it will solidify while cooling [4]. However the thermoset composite comes with variety of forms, not easily become softy on heating, never melt once set, reformed after cured and the process is irreversible [5]. In thermoset polymer, the composite fabrication process involved form of moulding, to shape the resin and reinforcement fiber.

2. The composite manufacturing process

Manufacturing process to fabricate the composite can be divided by two processes which are open and closed moulding. Matrix exposed atmosphere and surface condition cannot be control is called open mould, it consisted of hand layup, spray up and filament winding. Close mould is more practical to be called as liquid composite moulding (LCM) with this method fiber would be fill in both sides of mould and resin will add into the mould cavity within laminates, consist of vacuum infusion process, pultrusion, resin transfer moulding (RTM) and light resin transfer moulding (LRTM).

2.1 Open mould

2.1.1 Hand layup

Hand layup process using application as manual to lay down the single layers or the manually laying `plies' into a form of reinforcement known as `prepreg' [6]. Thousands of fibers pre-impregnated with resin by pouring, brushing, spraying, paint roller and bundled into tows with arrangement of fibers either single unidirectional ply or woven together. Ply manipulated into shape by using hand. Then it has to be firmly stuck to the mould surface or previous layer and leaving no air pocket between plies. Fiber which is woven, knitted, stitched or bonded fabrices which has been saturated with resin using hand [7]. This process include the roller to force the resin deep into the fiber and left the fiber till cure under standard atmospheric conditions. Material used for composite in this paper is glass fiber and epoxy resin. This process has been conducted in three times in different volume fraction 40:60, 50:50 and 60:40 with laminate size 300×300×3mm [8]. Then specimens that have been made were tested with four tests which are tensile testing, impact testing, shear testing and hardness testing, this test using glass fiber as the material and natural fiber (banana fiber) with s-glass fiber for composite [9,10]. This hand layup process is using epoxy and hardener 1:10 as it resin then cured around 6-8 hours [11]. The results have been found when fiber content was increased that will influence the increases of brittleness and delaminating in mechanical properties. Investigation shows that 50:50 is a better composition with less brittleness. This process is for low tooling cost, versatile range of products, easy to handle, however it required labour intensive, low volume process and time consuming. The product for large commercial parts

2.1.2 Spray layup

Spray layup process is spray simultaneously on the chopped fiber and resins into the mould by a pneumatic gun [12]. Bath tub fabrication using glass reinforced unsaturated polyster layer (GFRP) layers with separated resin (acrylic resin) by spray lay up method. The specimens must be moulded to obtain smooth surface before laminated by spray layup to get several colours and thermal stability. This thermal stability temperature must be consistent with resin during spray lay up process. Base and activator resin heated first with 80°C approximate 24 hours using band heater wrapper around barrels before operate spray layup process to find suitable viscosity during spraying work [13]. GFRP was chopped with unsaturated polyster resin through nozzles at the back side of the tube, to avoid it suffered from heat and water, then the tub surface is polished. Besides that, the utilization of GFRP with epoxy resin also can be use in spray system to improve the well-known spray technique and deliver good wettability [14]. Three point bending test ASTM D3039 from both directions was conducted to measure the acoustic emission (AE) on the specimens. Replacement on jute woven fabric reinforced unsaturated polyester layer (JFRP) from GFRP has achieved similar modulus and bending stiffness by thickness calculation. High bending modulus achieved with GFRP or JFRP. JFRP laminated reached 75% bending modulus with same thickness as GFRP. This showed that JFRP be able to replace the GFRP to fabricate bath tub. Jute fiber can be very high and deliver good elastic modulus [15]. This process able to be corrected if any errors happen and any material will replaced, however it has inconsistent process, unfriendly environment and longer curing time. The product is light structure such as bathtubs and small boats.

2.1.3 Filament winding

Filament winding mainly used for manufacturing open (cylinders) and closed end structures (pressure vessel or tanks) or open (cylinders) consists of winding filaments under tension over a rotating mandrel [18]. The mandrel will remove the hollow final product when the resin cured. Derivation on mandrel outline and spatial relation between the feed eyes [19]. Feed eyes on kinematic equations for coupling mandrel motion is actually based on analytic geometry. Design proposal on small-angle composite square tubes winding has develop winding trajectories non-slippage condition whether it accurate and realible. The presentation on winding theory regarding winding error has to be done to choose suitable design for composite square tubes. Winding angle for square tube produce accoding to the fiber pattern. The results shows to obtain uniform equation small angle the center angle must be 90° [20]. This angle needed to measure the mechanical strength [21]. Performance crashworthiness using filament winding gives effect on speed of crushing, material, treatment and structure of temperature including hybrid ratio, orientation of fiber and tube wall thickness capable to give energy absorption [22,23]. This process easy to be handle, can use the existing textile process and huge part size however it has mandrel which expensive to consult, facing part configuration with mandrel extraction. The product is to produce pipe, gas tank and aerospace parts.

2.2 Closed mould

2.2.1 Vacuum Infusion Process

Vacuum infusion process used vacuum pressure as to lead resin become a laminate. The application is before resin been introduced to dry material in mould [24]. After vacuum process complete, resin be sucked into laminate using appropriate placed tubing. The process aided by an assortment of materials and supplies. Technique of vacuum infusion is well known and established since 1950 [25]. The development of vacuum infusion in Delft university of technologies using a reactive thermoplastic polymer is called anionic polyamide-6 (APA-6) [26]. This development capable to manufacture textile fiber reinforced thermoplastic composites parts to become thicker, larger and more integrated. This fiber has been compacted to become fiber bed during vacuum application and mold filling stages [27,28]. Bagged fiber has to be rotated approximately 90° after applying vacuum and inserted it in a vertically placed heated platen press [29]. The infusion pressure to be compacted using pressure at 0.8 bar and not exceed 1.0 bar in 10 min [30]. Easier to fabricate fiber when it is compacted and it showed fill-times higher than resin transfer moulding [31]. This process used less wasted resin, low tooling cost and easy to be handled. However, the process is complicated more on fabricate large component.

2.2.2 Pultrusion

Process to shape polymeric composite materials into a part using constant cross section. This process used reinforcement fibers mat and placed on creel racks. The reinforcement fibers has pulled over guide plate, then soak it into resin bath. Then it goes to enter curing die with correctly shaped. This shape characterized to ensure the material easier to be inlet and capable to remove resin excess. For prevention on premature material solidification, water cooling channel is installed at first part of the die, heating platens placed at the bottom and top surfaces for heat material polymerization. Cured composite material is pulled with caterpillar outside of the die and travelling cut-off saw to cut the material to desired length [32]. This process used to fabricate fiber reinforced thermoplastic pre-impregnated materials as composites [33,34]. Taguchi or Design of Experiments methods used in this process to find full factor design which all levels combination of variables tested [35]. Comparison was made on expected theoretical prediction from rule of mixtures (ROM) with engineering conventional available materials. This ROM evaluate elastic modulus with accepting assumptions regarding material behaviour and series model [36]. Results on pultrusion process achieved in experimental modulus and theoretical found very good agreement. This process has economic way in terms of curing and impregnated which minimized fiber cost, however high cost in to constant the cross-section and has heat die. The products are pipe, gas tank and aerospace parts.

2.2.3 Vacuum Bagging

Process involved differential of acting cross two sides pressure of a flexible membrane to generate a uniformly distribution load. Uniform clamping force gives strong bonding since sheets bonding needs high surface level contact between mating surfaces [37]. Compression specimen and removed air from vacuum bag under atmospheric pressure was done by pumping or vacuuming to harden and compact process. Fiber mixture resin can become laminate, it happen by atmospheric pressure however balance of excess resin flow into catch pot. The clamping methods in this process used atmospheric pressure to grip resin-coated components from lamination until adhesive cures in room temperature [38,39,40]. Fiber inserted in single-sided mould to become fiber mat using resin with hand rollers. Silicone such as nylon film and sealant tape has been used to compact the wet laminate fiber using atmospheric pressure. Silicone is easy remove from mould and environmental friendly. It gives lesser time to fill resin flow and produce flexible thickness [41]. It also prevent high temperature occurred and compression for long periods of time over multiple cycles [42]. Vacuum and pump used to suck air inside mould into bleeder cloth and resin catch pot. This process are lower void content, higher fiber contain laminates and less amount volatiles emission, however it has additional cost for bagging material and required high operator skill especially on resin usage. The products are vehicles body part and door panels.

2.2.4 Resin Transfer Moulding (RTM)

RTM description process consisted of fiber that was closed in mould with mechanical clamp to avoid air moisture. Resin then injected into mould which has fiber with vacuum pressure until the fiber fully impregnated. Mould is open after the resin cured and composite removed from mould. Finishing operation on this process required to form a post-curing process. One inlet port used for resin injection and one outlet port is used to remove air inside the mould [43]. Evaluation on mechanical properties of fiber reinforced polyster has been investigated [44]. Injection pressure used in RTM about 200 kpa. Direction on fiber should be 0 and 90°. Preparation on resin (polyester) mixed with 1.5 % Methyl Ethyl Ketone Peroxide (MEKP) about 1.5% and cobalt (CoNAP) 0.2% cured the fiber in 30 minutes [45]. About four plies of fiber in different thickness have required, this thickness is not varying. However with small variation occurred give drastic impact on injection pressure [46]. Two injections constant with all parameters used to avoid repeatability, then water-jet cutting use to cut. Results gives 0 degrees is highest tensile strength compared others and increased volume fiber will increase tensile strength and flexural strength [47]. This process gives lower void content and high fiber volume good on safety environment, however it limited to smaller components and tooling cost so expensive. The products are vehicle seat and spar.

2.2.5 Light Resin Transfer Moulding (LRTM)

LRTM used rigid lower mould and semi-flexible, composite upper mould supported by steel. It is a hybrid combination between the wet bagging and vacuum compression moulding. Concept of LRTM is resin injected into the mould by an injection channel [48]. Inlet of the resin placed on the top of the mould at the same area with vacuum connection to control the amount of resin inside the mould. Fiber reinforcement is then placed into lower mould and sealed by the clamp. Preparation on carbon nanofiber (CNF)-loaded polyester (thermoset) composite can be in two different ways, injection or mix with gel-coat through mould [49]. This process put dry fiber reinforcement in mould and impregnated with thermoset liquid to be composite [50]. 2wt % CNF gel-coat showed volume resistivity below $10^5\Omega$ cm with 0.5 wt % CNF matrixes in electrical characterization. Polyester resin crystic used because low viscosity 250 Mpa.s and low exothermic with cobaltnaphthenate mixture methyl ethyl ketone peroxide (MEKP). The orientation of fiber could be random, 0^0, $40^0$and 90^0 respectively in three separate cavities [51]. LRTM Spartan II gives low pressure injection machine with two levels vacuum pump to assist resin flow and vacuum for mould closure [48]. The matrix cured in 24 hours room temperature and post-cured 40^0 for another 15 hours. Flexural properties tested according to ASTM D-790 and result showed flexural and tensile strength improved at 0.5 wt. % of CNF with better adhesion matrix CNF and fiber glass. Increasing CNF load will reduce electrical volume resistivity. This process has less cycle time, clean environment process and less wasted resin, however it tooling cost expensive and the process need to handle with care. The products are mold set, fittings and hoses.

3. Discussion

There are certain techniques to produce products using manufacturing process to fabricate composite. Concept to fabricate composite are open and closed mould. Open mould is suitable to produce low volumes part and cost cheaper than close mould. Other advantages of open mould are its effectiveness and used low density core material. Meanwhile for the closed mould concept more on to produce high volumes part with lower labour cost. Other benefit is to reduce waste and emission during the process and lead to clean environment as well as using less material and reduced disposal costs.

4. Conclusion

Development on fabrication composite depends on fiber and product selection for several applications. This open and closed mould concept provides an idea to produce high quality product. Combination and integration of each process has applied in order to obtain better results. Mechanical properties have been test at the end follow with ASTM standard to ensure the product and process capable to be used in manufacturing industry. Consideration on substitution of natural fiber to current synthetic materials should take into account for better opportunity in the future.

5. References

[1] Singh R 2006 Introduction to basic manufacturing processes and workshop technology *new delhi*:

[2] Nagavally R R 2016 Composite materials - history, types, fabrication techniques, advantages, and applications. *International Journal of Mechanical and Production Engineering*, **5**, 87-92.

[3] Agarwal B D, Broutman L J and Chandrahekhara K 2006 *Analysis and performance of fiber composite:*John Wiley & Sons.

[4] Grewall D and Benatar A 2007 Welding of plastics: fundamentals and new development, *International Polymer Processing Journal of the Polymer Processing Society*, **22**, 1-18.

[5] Capelot M, Montarnal D, Tournilhac F and Leibler L 2012 Metal-catalyzed transesterification for healing and assembling of thermosets. *Journal of the american chemical society*, **134**, 7664–7667.

[6] Elkington M, Bloom D, Ward C, Chatzimichali A and Potter K 2016 Hand layup: understanding the manual. *Advanced manufacturing: polymer & composites science,* **1**, 138-151.

[7] Swapnil S A, Sathesandip B, Chaudharibapu P and Vishal J 2017 Experimental investigation of mechanical properties of glass fibre/epoxy composites with variable volume fraction. *materials today: proceedings*, **4,** 9487–9490.

[8] El-wazerya M S, El-Elamy M I and Zoalfakarb S H 2017 Mechanical properties of glass fiber reinforced polyester composites.*International journal of applied science and engineering*, **14**, 121-131.

[9] Alam S, Habib F, Irfan M, Iqbal W and Khalid K 2010 Effect of orientation of glass fiber on mechanical properties of grp composites. *Journal of chemical association pakistan*, **32**, 265-269.

[10] Sakthivela R and Rajendran D 2014 Experimental investigation and analysis a mechanical properties of hybrid polymer composite plates. *International journal of engineering trends and technology* (ijett), **9**,1-8.

[11] Mutalikdesai S, Sujaykumar G, Raju A, Moses C J, Jose J and Lakshmanan V 2017 Mechanical characterization of epoxy/ basalt fiber/ flax fiber hybrid composites. *American journal of materials science*, **7**, 91-94.

[12] Xiao B, Yang Y, Wu X, Liao M, Nishida R, and Hamada H 2015 Hybrid laminated composites molded by spray lay-up process. *Fibers and polymers* ,**16**,1759-1765.

[13] Ha S K, Jang J G, Pang S H and Lee H K 2014 Advanced spray multiple layup process for quality control of sprayed frp composites used to retrofit concrete structures. *Journal of construction engineering and management*, **141**, 1-11.

[14] Ha S K, Jang J G, Pang S H and Lee H K 2013 An experimental study on sag-resistance ability and applicability of sprayed frp system on vertical and overhead concrete surfaces. *materials and structures* 1-13.

[15] Kikuchia T, Tani Y, Takai Y, Goto A and Hamadaa H 2014 Mechanical properties of jute composite by spray up fabrication method. *Eco-energy and materials science and engineering,* **11**, 289 – 297.

[16] Liakus J, Wang B, Cipra R and Siegmund T 2003 Processing–microstructure-property predictions for short fiber reinforced composite structures based on a spray deposition process. *Composite structures*, **61**, 363–374.

[17] Subham V 2015 Seminar on fiber glass composite. odisha, india: *Ajay binary institute of technology cuttek.*

[18] Shen F C 1995 A Filament-wound structure technology overview. *Materials chemistry and physics,* **42**, 96-100.

[19] Zu L, Xu H, Zhang B, Li D, Zi B and Zhang B 2018 Design and production of filament-wound composite square tubes. *Composite structures* 1-33.

[20] Fleischer J, and Schaedel J 2013 Joining automotive space frame structures by filament winding. *Journal of manufacturing science and technology*, **6**, 98–101.

[21] Abdalla F, Mutasher S, Khalid Y, Sapuan S and Hamoud B S 2007 Design and fabrication of low cost filament winding machine. *Materials and design* **28** 234–239.

[22] Xu J, Ma Y, Zhang Q, Sugahara T, Yang Y, and Hamada H 2016 Crashworthiness of carbon fiber hybrid composite tubes molded by filament winding. composite structures,**139**, 130–140.

[23] Sun G, Wang Z, Hong J, Song K and Li Q 2018 Experimental investigation of the quasi-static axial crushing behavior of filament-wound cfrp and aluminum/cfrp hybrid tubes. *composite structures*, 1-54.

[24] Pohl G and Pfalz M 2014 Innovative composite-fibre components in architecture. *Textiles polymers and composites for buildings,* 420–470.

[25] Anderson M H, Lundstrom S T and Gebart R B 2016 Numerical model for vacuum infusion manufacturing of polymer composites. *International journal of numerical methods for heat & fluid*, **13**, 383 - 394.

[26] Rijswijk K V, Teuwen J J E, Bersee H E N and Beukers A 2009 Textile fiber-reinforced anionic polyamide-6 composites. part i: the vacuum infusion process. *Composites: part,* **40,** 1-10.

[27] Yenilmez B and Sozer E M 2009 Compaction of e-glass fabric preforms in the vacuum infusion process a: characterization experiments. *Composites: part* **40**, 1458–1464.

[28] Yenilmez B, Senan M and Sozer E M 2009 Variation of part thickness and compaction pressure in vacuum infusion process. *Composites science and technology*, **69**,1710–1719.

[29] Andersson H M, Lundstrom T S, Gebart B R and Langstrom R 2002 Flow-enhancing layers in the vacuum infusion process. *Polymer composite*, **23**, 895-901.

[30] Hamami A and Gebart B R 2000 Analysis of the vacuum infusion molding process. *polymer composites,* **21**, 28-40.

[31] Modi D, Johnson M, Long A and Rudd C 2009 Analysis of pressure profile and flow progression in the vacuum infusion process. *Composites science and technology* **69,** 1458–1464.

[32] Carlone P, Palazzo G S and Pasquino R 2006 Pultrusion manufacturing process development by computational modelling and methods *Mathematical and computer modelling* **44**, 701–709.

[33] Novo P J, Silva J F, Nunes J P and Marques A A 2015 Pultrusion of fibre reinforced thermoplastic pre-impregnated materials. *Composites part* b, 1-21.

[34] Memon A and Nakai A 2013 Mechanical properties of jute spun yarn/pla tubular braided composite by pultrusion molding. *Energy procedia* **34**, 818 – 829.

[35] Carlsson A and Astrom T B 1998 Experimental investigation of pultrusion of glass fibre reinforced polypropylene composites. *Composites part a* **29**, 585-593 .

[36] Oprişan G, Taranu N, Mihai P, Popoaei S and Enţuc I S 2013 Stiffness properties along principal material axes of fibre reinforced polyester pultruded plate elements 1-10.

[37] Cassano C L, Simon A J, Liu W, Fredrickson C and Fan Z H 2014 Use of vacuum bagging for fabricating thermoplastic microfluidic devices. *Royal society of academy*, 62-66.

[38] Krishnaiah M, Reddy D M and Kumar A K 2016 Fabrication of silicone bagging and fluid flow analysis in vacuum bag moulding process. *International journal of engineering science and computing,***6**,2793-2798.

[39] Hassan M H, Othman A R, Abdullah J and Mahmud A S 2016 Effect of bagging configurations on vacuum bagging only-oven cured to the thickness variations for the comple-shaped laminate composite. *Journal of scientific research and development*, **3,** 41-46.

[40] Hartono R M and Subawi H 2016 Optimization of the composite repair work using 5h satin dry glass fabric and epoxy resin ly5052/hy5052 materials through the vacuum bagging technique. journal of physics: *Conference series*, **739**,1-6.

[41] Harper A 2016 Reusable vacuum membranes do save money. *Reinforced plastics* **60**, 380-386.

[42] Malek P and Walczyk D 2016 Silicone rubber properties during consolidation/curing of advanced composites using specialized elastomeric tooling. *Journal of manufacturing science and engineering,* **138**, 1-7

[43] Robertson F C 1988 Resin transfer moulding of aerospace resins-a review. *British polymer journal* **20**, 417-429.

[44] Santa J F, Jaramillo J and Patiño I D 2016 Mechanical characterization of composites manufactured by rtm process: effect of fiber content, strain rate and orientation. *Latin america journal of solid and structures*, **13**, 344-364.

[45] Sreekumar P A, Saite J M, Joseph K, Unnikrishnan G and Thomas S 2012 Electrical properties of short sisal fiber reinforced polyester composites fabricated by resin transfer molding. *Composites: part a,* **43**, 507–511.

[46] Olivero K A, Hamidi Y K, Aktas L and Altan M C 2003 Effect of preform thickness and volume fraction on injection pressure and mechanical properties of resin transfer molded composites. *Journal of composite materials,* **38**, 937-957.

[47] Wang B C, Zhou X and Ma K M 2013 Fabrication and properties of cnts/carbon fabric hybrid multiscale composites processed via resin transfer molding technique. *composites: part b*, **46**, 123-129.

[48] Isoldi L A, Souza J A, Santos E D and Marchesini R 2013 Constructal Design Applied To The Light Resin Transfernoulding (Lrtm) Manufacturing Process. *22nd International Congress Of Mechanical Engineering*, **22**, 4845-4855.

[49] Morales G, Barrena M, Salazar J G, Merino C and Rodríguez D 2010 Conductive cnf-reinforced hybrid composites by injection moulding . *Composite structures,* **92**, 1416–1422.

[50] Timms J, Bickerton S and Kelly P 2016 Numerical simulation of the rtm light manufacturing process. *International conference on composite materials* **18,** 1-6.

[51] Hutchinson J R and Schubel P J 2014 A coupled structural and flow approach for numerical simulation of the light resin transfer moulding process. ii: fabric permeability and compaction characterisation, model results and a 6-kw wind turbine blade case study. *Journal of reinforced plastics and composites,* **33**, 1237–1257.

IOP Conf. Series: Materials Science and Engineering **488** (2019) 012009 doi:10.1088/1757-899X/488/1/012009

Experimental testing of oil palm fibre composite manufactured via vacuum bagging method

N S Binti Mohd Hafidz, M S Bin Mohamed Rehan

Department of Mechanical Engineering, Kulliyyah of Engineering, International Islamic University Malaysia, Malaysia

nurulsyazwanihafidz@gmail.com

Abstract. The objective of this research is to compare tensile strength of oil palm fibre composite produced from oil palm fibre of different configurations. Compressed random oil palm fibre and uncompressed random oil palm fibre reinforced polyester composite were manufactured using vacuum bagging technique. The volume fractions of the fibre to the resin were 6.6:93.4 and 17:83 for compressed fibre and uncompressed fibre composites, respectively. The ratio of the polyester to hardener is 100:2. Tensile properties obtained via tensile tests as per ASTM D 638 specifications using Universal Testing machine INSTRON 5582. Load-displacement plots show that the fracture load and the slope for specimens from the same plate vary with a significant range. Stress-strain plots also have the same issue. With such data direct comparison of average value may not be giving a conclusive comparison, with a big standard deviation for both sets of specimens. Furthermore, there are two simultaneous factors in this study, namely the configuration of the oil palm fibres used and the volume ratio of the composite. Further testing and analysis is deemed to be necessary to further fine tune the fabrication process, for a conclusive comparison.

1. Introduction

1.1 Background

Even back in 2006 natural fibre composites was reported to have been embraced for a decade by European car manufacturers and suppliers for door panels, seat backs, headliners, package trays, dashboards and interior parts [1]. While stating natural fibres do offer such benefits as reduction of weight and cost, it was also stated that several major technical considerations must be addressed before a wide scale acceptance, particularly in exterior parts where a Class A surface finish is required. Among the challenges mentioned were the homogenization of the fibre's properties and a full understanding of adhesion between the fibre and matrix.

There has been a great development in natural fibre composites, and a lot of studies have been done on their mechanical performance [2]. The focus for this study will be in the application of oil palm empty fruit bunch (OPEFB) as reinforced material in composite materials. A review by Mahjoub et. al. [3] focusing on the properties of OPEFB fibre and mechanical properties of OPEFB composites in term of tensile and flexural properties highlighted that previous studies show the impact properties of OPEFB composites are better than pure resin that it can be interesting factor for special industries for using new material to make products such car bumper and dashboard. However, it was also noted that more studies are needed to find out and verify the predictions. A review study on the potential of OPEFB as reinforcing materials in polymer composites for energy absorption during low-velocity impact by Faizi et. al. [4] suggests high potential applications of OPEFB as reinforcing materials in composite structures.

Mohamed Yusoff et al [5] have studied the mechanical properties of short random oil palm fibre reinforced epoxy composites, where empty fruit bunch was selected as the fibre. The composites were fabricated by hand lay-up techniques, and composite plates of four different volume fractions were fabricated. The study observed that the tensile and flexural properties showed a decreasing trend as the fibre content increased.

Al-Oqla and Sapuan [6] have published their work to analyze the feasibility of date palm fibres for sustainable automotive industry. In this work they stated to have successfully able to categorize and tabulate the criteria that affect natural fibre reinforced polymer composite materials in distinguished levels for the first time. For date palm fibre they found that they have very competitive properties suitable for automotive industry.

Sahari and Maleque [7] have investigated the mechanical properties of oil palm shell composites with different volume fraction using unsaturated polyester as matrix. The study found that the tensile strength of the composite increased with oil palm shell content. The study also mentioned that SEM analysis indicates that the filler detached from the fracture surface which proved the poor filler-matrix interfacial bonding.

Jawaid et al [9] have studied the effect of addition of jute fibres to oil palm composite and found that it increases the storage modulus. The overall use of hybrid system was found to be effective in increasing tensile and dynamic mechanical properties of the oil palm-epoxy composite probably due to the enhanced fibre/matrix interface bonding. They did mention that while many automotive manufacturers try to replace synthetic fibres with natural fibres, it is not comparable in properties, and stated that fabricating hybrid composites by the combination of two natural fibres gives them advantages to replace synthetic fibres.

1.2 Mechanical Properties of Oil Palm Fibre

Mahjoub et. al. [3] have compiled the fibre properties of oil palm empty fruit bunch fibre gathered from various researchers, and reported results are not identical because of variation in the kind of oil palm fibre used. Additional factor of irregular sectional area was also mentioned [10]. Consequently, some of the references reported wide range of value for mechanical properties.

1.3 Tensile Properties of Oil Palm Fibre Composites

Tensile strength and tensile modulus are considered as major properties of materials by engineers. Mahjoub et. al. [3] have compiled the tensile properties of oil palm composites obtained from various researchers. They have noted that the compiled results indicate that adding oil palm fibre to polymer causes the reduction of tensile strength of pure resin.

2. Study Methodology

In this study two types of oil palm fibre were used, which were compressed random oil palm fibre and uncompressed random oil palm fibre. There were four stages in doing this research, which are the compressed oil palm fibre and uncompressed oil palm fibre preparation, oil palm fibre reinforced polyester composite fabrication process, cutting process of the composites by using CNC milling machine and the last stage was tensile test was done. It was to obtain mechanical characteristics of oil palm fibre reinforced polyester composite.

2.1 Fibre material

Untreated oil palm fibre is obtained from Malaysia Palm Oil Board in Bangi, Selangor, Malaysia. The raw fibre is obtained in two types as shown in Figure 1, where the first one is random uncompressed fibre, and the second is compressed into a mat of certain roughly uniform thickness. The uncompressed random fibre obtained in bundles, tangled and intermixed between its fibres. In order to prepare the uncompressed random fibre, the fibre needed to be combed so that it would produce untangled and more uniformly distributed fibre. On the other hand, compressed random fibre obtained is very compact and dense in nature.

(a) (b)

Figure 1. Oil palm fibre obtained in the form of (a) uncompressed random fibre and (b) compressed random fibre.

2.2 Fabrication of oil palm fibre composites

The oil palm fibre reinforced polyester composites were manufactured using vacuum bagging fabrication technique. The fabrication technique was same for both types of oil palm fibres, where some of the main steps are shown in Figure 2.

(a) (b) (c)

Figure 2. Fabrication process of the oil palm fibre composite plates; (a) mixed polyester resin and hardener was impregnated through the layer of the fibre, (b) vacuum pump draws out the air inside the confined space, and (c) sample oil palm fibre composite plate.

A 320 mm x 220 mm aluminium tray was used as a mould to fabricate the composites. A layer of mylar sheet was placed on each side of the plate as a releasing agent. Polyester resin and hardener solution were mixed thoroughly according to specific fibre-matrix volume fraction ratio, as shown in table 1. The mixed solution is then poured into the tray until it fully filled the whole tray. The fibre was then placed in the tray and soaked it thoroughly until all the fibres are covered with resin. The mixed solution was then impregnated through the layer of fibres. Vacuum bagging materials was placed on top of the wet fibres. Vacuum bagging film enclosed the lay-up to provide a confined space which is sealed using sealant tape. A vacuum pump draws out the air inside the confined space to produce an airtight vacuum bag until curing process is done.

IOP Conf. Series: Materials Science and Engineering **488** (2019) 012009 doi:10.1088/1757-899X/488/1/012009

Table 1. Uncompressed Random Fibre and Compressed Random Fibre Composites Specimen Parameters.

Parameter	Uncompressed Random Fibre	Compressed Random Fibre
Weight of fibre, W_f	15.3	47.5
Volume Ratio	(6.6/93.4)	(17/83)
Mass Ratio	(15.3/424.8)	(47.5/453.8)
Test Specimen	R1 R2 R3 R4	M1 M2 M3 M4
Weight of composite, $W_c = W_f + W_r$	440.1 g	501.3 g

2.3 Cutting of specimens
There were two different plates which were uncompressed random fibre plate and compressed random fibre plate. Each plate was cut into four specimens. Hence, the total of the specimens were eight. Tensile test specimen was prepared according to the standard size and specification of ASTM D 638 using CNC milling. Table 2 shows the dimensions for each of the plate.

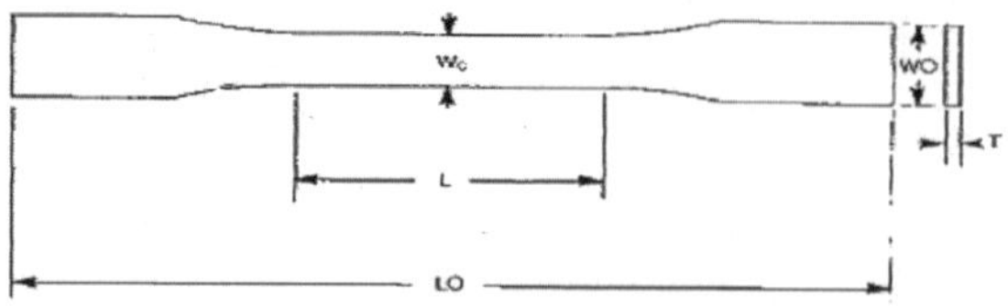

Figure 3. Specimen Dimensions.

Table 2. Specimen Dimensions.

Dimensions (mm)	Uncompressed Random Fibre	Compressed Random Fibre
WO	19	19
L	57	57
LO	165	165
W_c	13	13
T	5	6.5

2.4 Tensile test
Tensile strength and modulus of elasticity for uncompressed random fibre and compressed random fibre reinforced polyester composite were tested by using INSTRON 5582 Ultimate Testing Machine. This measurement was conducted according to ASTM D 638 procedures at a test speed of 5mm/min. Four specimens were tested for each case and its average value is tabulated accordingly.

3. Results and Discussion
3.1 Load vs. Displacement Plot
The plots of load against displacement are as presented in Figure together with the images of the fractured region. One major contrast between specimens fabricated from uncompressed random fibre and from compressed random fibre is the specimens from uncompressed fibre tend to break into two separate pieces when they fractured, while for specimens from compressed fibre the fracture across the width of the specimen can be seen, but the specimen still is intact, illustrating that the compressed fibre is not totally cut separated when the fracture occurs. This may be due to the fact that the compressed

fibre specimens have the fibres tangled with one another in a very compressed manner, as compared to the uncompressed fibre where the individual fibre can be cut off more easily when fracture occurs.

Another observation that can be made is that the specimens from the same fibre type do have a big range of both fracture load and the slope of load-displacement curve. Based upon visual inspection, it is quite unclear what the factor behind the difference of fracture load is. It can be argued that the random nature of the alignment of the fibre throughout the length of the specimen and between specimens inevitably lead towards the difficulty to have an exact uniform behaviour between specimens.

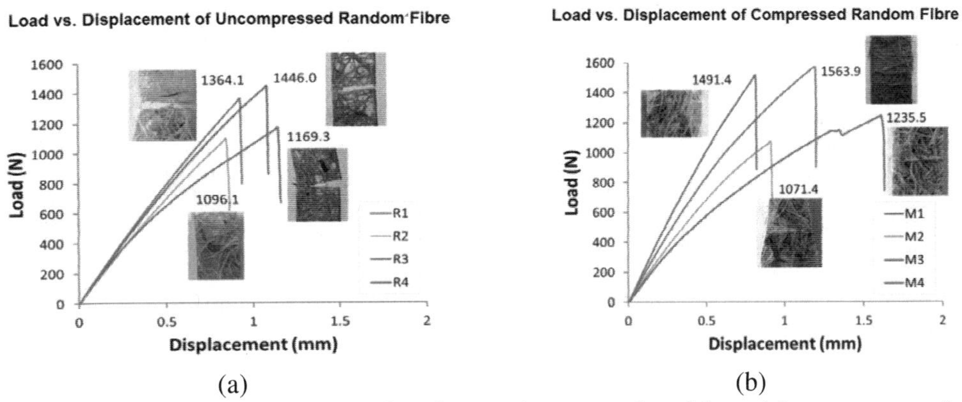

Figure 4. Load vs. Displacement plots for specimens produced from (a) uncompressed random fibre and (b) compressed random fibre.

3.2 Stress vs. Strain Plot

The plots of stress against strain are as shown in Figure 5. The range of the values of slopes between specimens of the same fibre type is also quite big. With this observation it is a big question whether oil palm composite plates like these can have a material property or it will be diverse throughout the plate due to the random characteristic of the fibre.

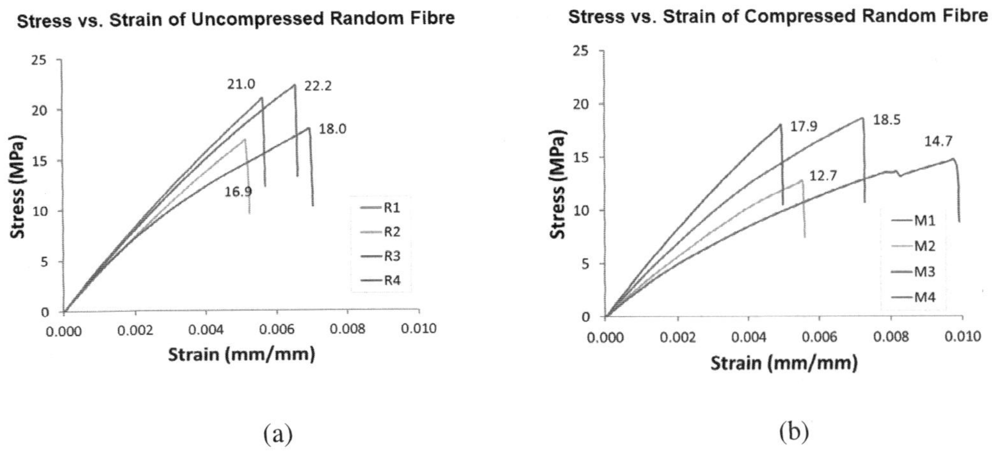

Figure 5. Stress vs. Strain plots for specimens produced from (a) uncompressed random fibre and (b) compressed random fibre.

IOP Conf. Series: Materials Science and Engineering **488** (2019) 012009 doi:10.1088/1757-899X/488/1/012009

3.3 Modulus of elasticity

From the graph stress vs. strain as shown in figure 5, modulus of elasticity can be obtained. One aspect that can be observed from the plots shown in figure 5 is the fact that it is difficult to distinguish the linear elastic region for each curve. This has a direct impact on the calculation of tensile modulus. Hence, the calculation of modulus is based on the slope of the tangent to the stress-strain curve at a low stress. Based on table 3, it shows the average tensile modulus for uncompressed random fibre composite and compressed random fibre composite. Figure 6 shows more clearly the average tensile modulus in a bar graph.

Table 3. Modulus of elasticity and standard deviation for uncompressed random fibre and compressed random fibre.

Parameter	Uncompressed Random Fibre	Compressed Random Fibre
Average Tensile Modulus (MPa)	4171.98	3476.63
Standard Deviation	221.072	705.259

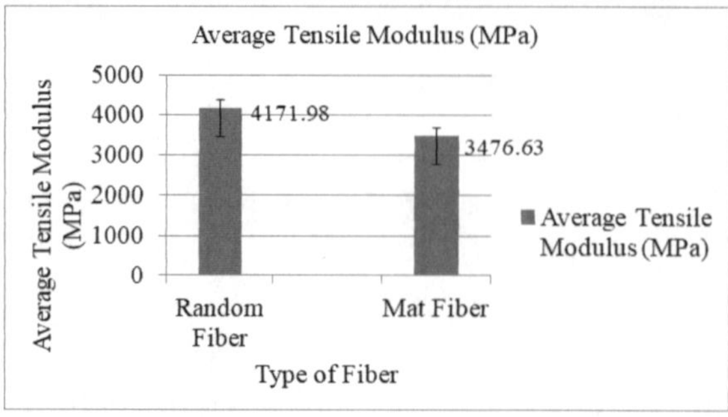

Figure 6: Average Tensile Modulus.

With these observations, it is deemed important to further analyse the fabrication process as well as the composition of the oil palm fibres in the composite. The future analysis will also benchmark the tensile behaviour to that of pure polyester and also glass fibre composite, to analyse its potential in term of its application.

4. Conclusion

Two types of oil palm fibres have been used to fabricate specimens of oil palm fibre reinforced polyester composite. Tensile tests were performed to analyse the tensile mechanical properties between the two sets of composite. Load-displacement plots show that the fracture load and the slope for specimens from the same plate vary with a significant range. Stress-strain plots also have the same issue. With such data direct comparison of average value may not be giving a conclusive comparison, with a big standard deviation for both sets of specimens. Furthermore, there are two simultaneous factors in this study, namely the configuration of the oil palm fibres used and the volume ratio of the composite. Further testing and analysis is deemed to be necessary to further fine tune the fabrication process, for a conclusive comparison.

References

[1] Holbery J and Houston D 2006 Natural-fibre-reinforced polymer composites in automotive applications 2006 *J. Miner. Met. Mater. Soc.* **58** 80-86

[2] Pickering K L, Efendy M G A and Le T M 2016 A review of recent developments in natural fibre composites and their mechanical performance 2016 *Compos. Part A Appl. Sci. Manuf.* **83** 98-112

[3] Mahjoub R, Yatim J M and Sam A R M 2013 A review of structural performance of oil palm empty fruit bunch fiber in polymer composites 2013 *Advances in Materials Science and Engineering*. **2013** 9

[4] Faizi M K, Shahriman A B, Majid M S A, Shamsul B M T, Ng Y G, Basah S N, Cheng E M, Afendi M, Zuradzman M R, Wan K and Hazry D 2017 An overview of the oil palm empty fruit bunch (OPEFB) potential as reinforcing fibre in polymer composite for energy absorption applications 2017 *MATEC Web Conf.* **2016** 1-9

[5] Yusoff M, Salit M S, Ismail N and Wirawan R 2010 Mechanical properties of short random oil palm fibre reinforced epoxy composites 2010 *Sains Malaysiana.* **39** 87-92

[6] Al-Oqla F M and Sapuan S M 2014 Natural fiber reinforced polymer composites in industrial applications: feasibility of date palm fibers for sustainable automotive industry 2014 *J. Clean. Prod.*

[7] Sahari J and Maleque M A 2016 Mechanical properties of oil palm shell composites 2016 *Int. J. Polym. Sci.* **2016** 7

[8] Jawaid M, Khalil H P S A, Bakar A and Khanam P N 2011 Chemical resistance, void content and tensile properties of oil palm/jute fibre reinforced polymer hybrid composites 2011 *Mater. Des.* **2** 1014-1019

[9] Jawaid M, Khalil H P S A, Hassan A, Dungani R and Hadiyane A 2013 Effect of jute fibre loading on tensile and dynamic mechanical properties of oil palm epoxy composites 2013 *Compos. Part B Eng.* **2012** 6

[10] Virk A S, Hall W and Summerscales J 2010 Failure strain as the key design criterion for fracture of natural fibre composites 2010 *Compos. Sci. Technol* **70** 995–999

Experimental study on low temperature power generator

S I Ihsan*, AKM Mohiuddin, S N Syed Abu Bakar and S S Rosili

Department of Mechanical Engineering, Kulliyyah of Engineering, International Islamic University Malaysia

*E-mail: sihsan@iium.edu.my

Abstract. An experimental set-up for Steam Rankine Cycle (SRC) is designed and fabricated by using affordable materials and components. SRC is a promising system for low temperature power generation application. Experiments are then conducted to analyse the effect of working fluid volume and boiler pressure on system performance. Results indicate agreeable relationship of the variables and system performance but with relatively low overall efficiency due to design and fabrication limitations.

Keywords: Low Temperature Power Generation, Low Grade Heat, Steam Rankine Cycle

1. Introduction

People around the globe nowadays are busy conferring about the greenhouse effect that is getting worse day by day. Greenhouse effect is basically good to the planet Earth when the greenhouse gases is capable of trapping and absorbing the radiation sent by the Sun in order to keep this planet warm for humans. However, when there is too much of greenhouse gases like Carbon Dioxide and Methane, the Earth's surface temperature increase which then lead to global warming. They believe that human activities have been one of the major causes for this circumstance to come up. Every day, every second, everywhere, there is waste heat emitted from the industrial works; unused and unrecovered, exhausting to the atmosphere.

This waste heat released by industry can actually be classified as low grade heat which is in a temperature range of 100°C to below 150°C. This low grade heat not only comes from industrial process, but has also been produced by biomass energy, solar energy, parabolic trough or turbine exhaust heat. There are a lot of potentials from this waste heat that some researchers have already discovered. For example, Siemens researchers in Erlangen have developed a demonstration plant by using waste heat in order to vaporise wastewater within the temperature range of 70°C to 120°C. This group of researchers have also built a heat pump that can raised temperature up to 140°C to be used to boost the low temperature of waste heat from 70°C to 130°C in order to utilize the heat to warm a building [1].

More than that, this low grade heat actually has a big potential in generating electrical power by using special systems. Thus, the main concern that will be discussed in this study is regarding the systems that can be used to generate electricity by using low grade heat sources. There are few number of systems that can be employed to generate electrical power by using low grade heat sources such as Rankine Cycle, Kalina Cycle and Stirling Cycle. Rankine Cycle is an ideal vapour power cycle that is widely and commonly used in steam power plants to generate electrical power in the world. Rankine Cycle make use of the characteristics of water that is easily available, low cost and renewable as its working fluid. Apart of that, Organic Rankine Cycle (ORC) is a "modified" Steam Rankine Cycle that replace steam with organic working fluids. ORC uses a single substance as working fluid like 100% ammonia, sulphur dioxide or isopentane.[2-4].

Rankine Cycle was invented by W. Rankine in 1850s and is widely used for boiler steam power generation. A traditional Rankine Cycle is called as Steam Rankine Cycle where it uses water as its working fluid. Rankine Cycle is an ideal cycle for vapour power cycle and it is introduced to overcome some limitations and difficulties in Carnot Cycle in operating a vapour power engine.

In this project, an experimental set-up for steam Rankine Cycle (SRC) was designed in order to conduct a steam cycle experiment. The data will then be calculated and analyzed to compute the efficiency. Water has several desirable characteristics like high availability, renewable and low cost. It is believed to be more worthy for it to be used in huge portion on a system compared to the usage of high cost refrigerants in the market; especially to know that there is no ideal working fluid because they all have their own pros and cons. On top of that, this designed system that only use water as its working fluid can also be used in rural area, where there is no access of electricity, to generate electricity since source of water is wider than the source of certain working fluids.

2. Design and Methodology

2.1. Steam Rankine Cycle (SRC)

Rankine Cycle consists of Hydraulic Working Pump, Evaporator, Turbine and Condenser as shown in figure 1. Firstly, the working fluid is sent to the Evaporator from Hydraulic Working Pump where it evaporates this working fluid to become vapour through heat exchanging process with the hot heat source. The vapour is then sent to the Turbine and produces work by generating electricity. The exhaust vapour from the Turbine is condensed to water by low heat source (cooling water) through heat exchanger. Lastly, the condensate is sent back by the Hydraulic Working Pump to the Evaporator and repeat the cycle. Therefore, since it is called as Steam Rankine Cycle, the working fluid used is water which can be changed to steam at high temperature by applying heat.

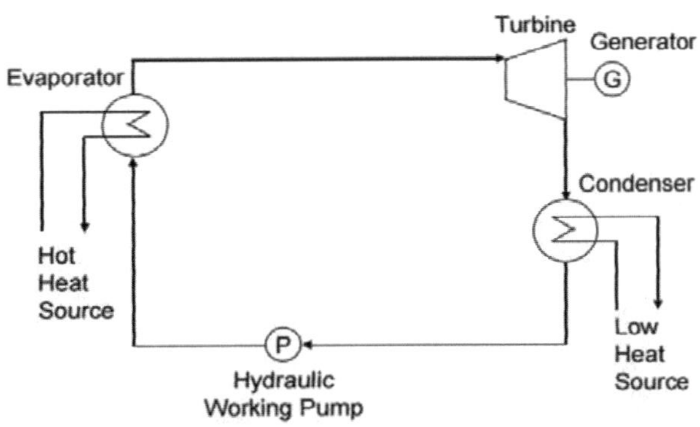

Figure 1. Rankine Cycle. Source [5].

2.2. Details of Experimental Setup

Figure 2 basically represent the schematic diagram of the experimental set-up for SRC. It consists of Working Pump, 2 Heat Exchangers that function as Evaporator in boiler and Condenser respectively, Turbine with generator to produce electrical power and also connecting pipes. The working pump is important in circulating the working fluids back from Condenser to the Evaporator. Meanwhile, connecting pipes are used to connect one component to each other. The working fluids for the system is water.

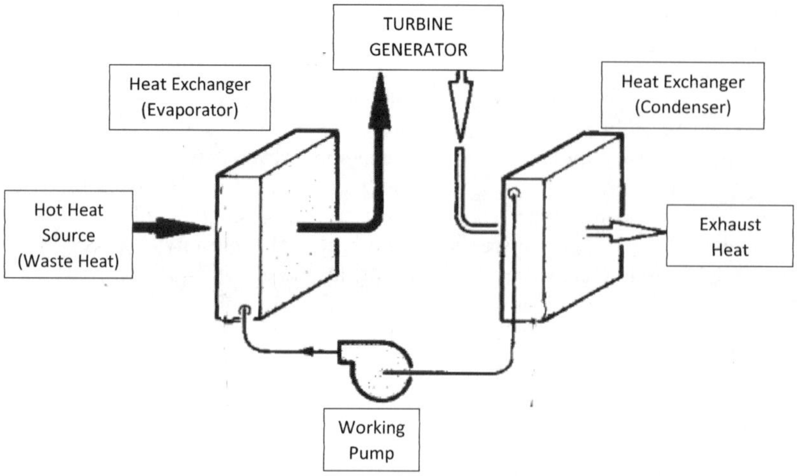

Figure 2. Schematic Diagram of Steam Rankine Cycle Experimental Set-Up.

Main components of the system will be briefly discussed.

2.2.1. Boiler

As shown in figure 3, boiler is made up of 5.5 inches diameter of mild-steel pipe with 16 inches length which is closed at both ends (welded with circle-shaped pieces of mild-steel). About 1 inch from the bottom of the boiler, there is a 0.5 inch diameter pipe located horizontally inside the boiler to provide a space where a gas torch can be inserted inside the boiler for the water heating purpose.

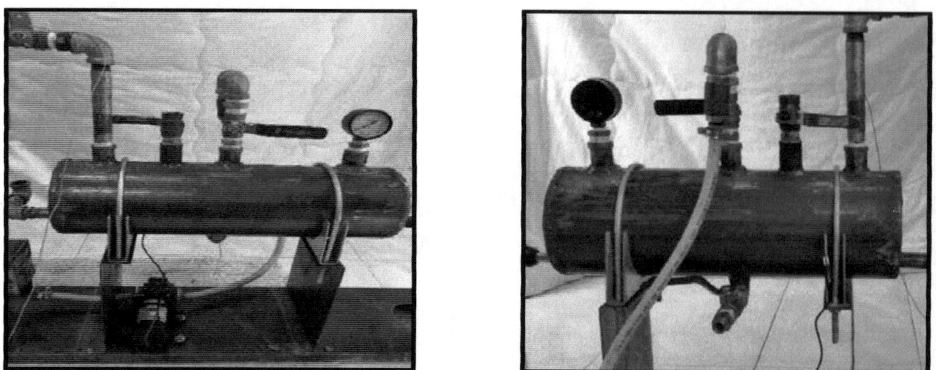

(a) (b)

Figure 3. Boiler front view (a) and rear view (b).

At the bottom of the boiler, there is a ball valve that acts as an outlet pipe for water out. There is another ball valve used for water inlet at the top of it. Meanwhile, at the centre top of the boiler is a GI socket attaches with a hose that connects the boiler with the water pump. A pressure gauge is also located at the right top of the boiler for measurement purpose, in order to check for the pressure inside the boiler throughout the experiment. Whilst at the left top of the boiler, there is a GI pipe to connect the boiler to turbine. This 2 Litre capacity boiler is perfectly placed on the boiler clamps which used to hold the boiler tight and secure at a certain height since the torch from gas tank will be inserted into the hole made on the side of the boiler for the heating process.

2.2.2. Turbine

Turbine is made up of eight mild steel rods of 4.5 inches length with small square-shaped blades riveted at the end of each rods. Located at centre of the turbine is a 1 inch diameter of circle plate which is attached with these 8 rods; divided evenly with the same distance and angle. It is designed that way in order to reduce the load that the steam needs to push in order to rotate the blades. This turbine is placed inside a turbine cover that has 2 openings. The first opening is where the steam nozzle connected boiler took place while the other one connects the turbine cover with condenser tube that has fins. This is shown in figure 4.

(a) (b)

Figure 4. Turbine blade view (a) and complete with cover (b).

2.2.3. Condenser

Condenser is divided into 2 sections; the first section is the one that connects from turbine cover which is the condenser tube (with fins). Meanwhile, the second part is the condenser tank which sandwiches between the condenser tube and water pump. The condenser tube is 1.5 inch hollow in size with 14 fins made up of mild steel plates attach to it. The size of each fin is 2.8 x 2.8 inch. On the other hand, for the condenser tank, it is made up of 12 x 3 x 3.3 inches of stainless steel rectangular-shaped box. This tank is used to collect the condensate coming from the condenser tube. The actual fabricated condenser is shown in figure 5.

(a) (b)

Figure 5. Condenser tube with fins (a) and condenser tank (b).

2.2.4. Other components

Water pump is located in between the condenser tank and boiler. A 12V, 3.9 litre per minutes (LPM), 80 psi pump is used and connected to a 12V battery. DC motor is connected to the turbine as load to measure the voltage output produced by the turbine. A 12V brush motor is used. Connecting pipes are made up of GI pipes that include elbow joint, steam valve and steam nozzle.

2.2.5. Measuring devices

A digital multi meter is used in order to measure the voltage output readings of the DC motor. Meanwhile, for pressure reading inside the boiler and condenser, two 100 psi pressure gauges were placed at these two respective points. For temperature readings at the inlet and outlet turbine, electronics waterproof temperature sensors are used. Once the sensors detect the temperature difference inside the system, display will be shown on PC screen in ARDUINO Software that is already programmed and coded to these two sensors.

At the turbine inlet, 1 meter K-Type thermocouple with amplifier breakout v2.0 MAX31855 is located while DS18B20 black waterproof temperature sensor is placed at the turbine outlet. K-Type thermocouple with amplifier breakout v2.0 MAX31855 is designed to detect higher temperature, specifically more than 100°C while DS18B20 black waterproof temperature sensor is suitable for below 125°C. These two temperature sensors were connected to ARDUINO board to synchronise the readings using ARDUINO Software installed in laptop.

Based on the component design, a lab size prototype system was fabricated as shown in figures 6 and 7.

Figure 6. Steam Rankine cycle system's front view.

Figure 7: Steam Rankine cycle system's back view.

Two different experiments were conducted:

1. Identifying the effect of different volume of working fluid used on system performance
2. Identifying the effect of different boiler pressure used on system performance.

In the first experiment, the boiler is filled with the specified amount of water, heat is supplied to the boiler and the voltage output is measured at the turbine. For the second experiment, 1 L of water is filled in the boiler and heat is supplied to the boiler. Once the pressure in the boiler has reached the specified value, the valve is open to let it pass through the turbine. Voltage output is measured. The variation of working fluid volume and boiler pressure is summarized in Table 1 below.

Table 1. Variables in Both Experiments.

	Exp 1: Working Fluid Volume (L)	Exp. 2: Boiler Pressure (MPa)
1	0.75	0.1
2	1.0	0.2
3	1.5	0.34

3. Results and Discussions

As mentioned earlier, there are two parts of the experimental work, namely to analyse the effect of working fluid volume and the effect of pressure in boiler.

3.1. Effect of Working Fluid Volume Change

Figure 8 shows that 0.75L of water produces the least voltage output which is 1.2V while 1.5L of water produces the largest voltage which is 1.8V. Therefore, the higher the volume of water inside the boiler, the higher the voltage produced by the motor. Big amount of water produces a big amount of steam that will generate a huge force to push the turbine to generate more work, producing higher voltage as a result.

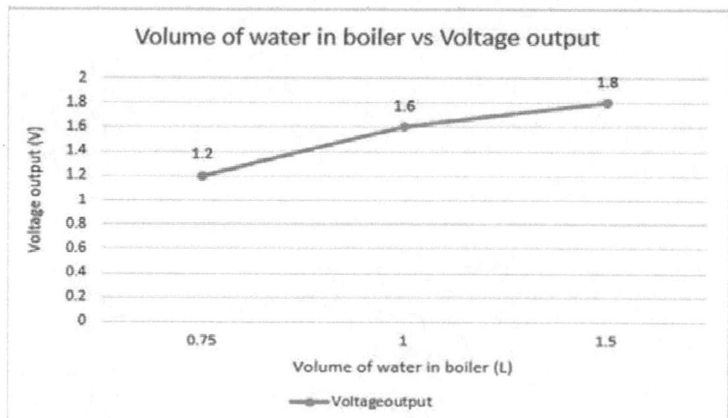

Figure 8: Volume of water in the boiler versus voltage output.

On the contrary, figure 9 shows that 0.75L of water has the highest thermal efficiency, as much as 23.11% and followed by 17.3% efficiency produced by 1L of water while 1.5L of water gives the lowest efficiency which is only 15%. Thermal efficiency is the ratio between the work done by the system and energy added to the system. Hence, by analysing amount of energy added and rejected on figure 10, it can be inferred that the highest volume of water produces the highest heat losses to the surrounding which indicates the low efficiency.

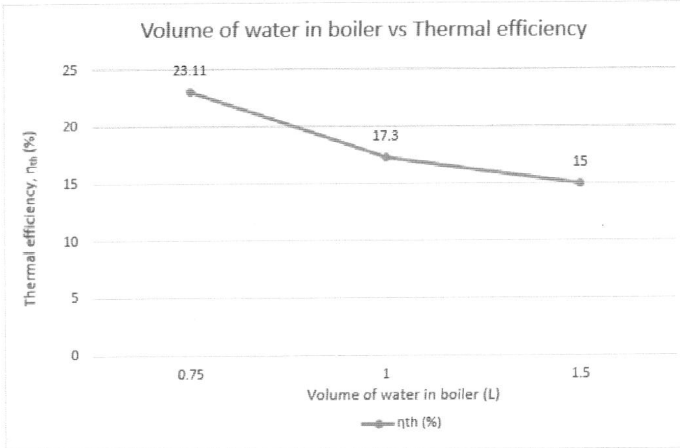

Figure 9: Volume of water in the boiler versus thermal efficiency.

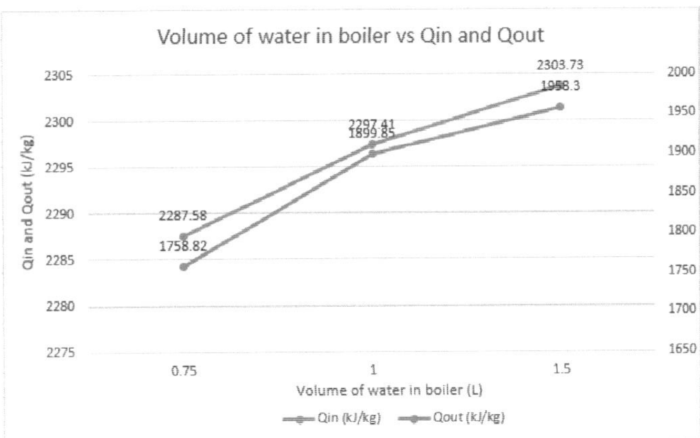

Figure 10: Volume of water in the boiler versus energy added in the boiler and energy rejected in the condenser.

3.2. Effect of Pressure Change

Figure 11 shows that 0.1MPa pressure of water produces the least voltage output which is 1.13V while 0.34MPa pressure of water produces the largest voltage which is 2.05V. This indicates that the higher the pressure of water inside the boiler, the higher the voltage produced by the motor, as expected according to thermodynamics.

IOP Conf. Series: Materials Science and Engineering **488** (2019) 012010 doi:10.1088/1757-899X/488/1/012010

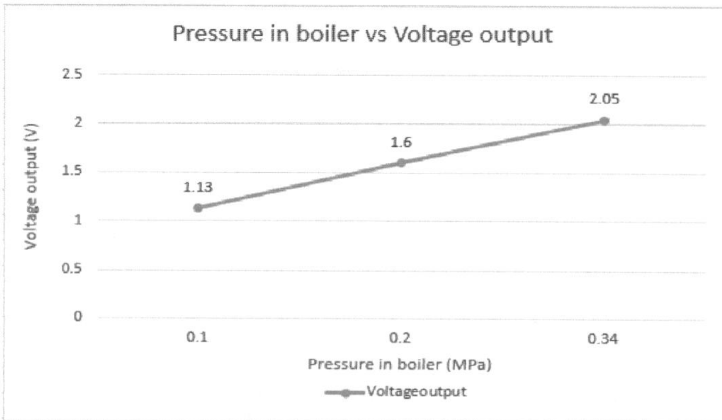

Figure 11: Pressure in the boiler versus voltage output.

Figure 12 shows that 0.2MPa pressure of water has the highest thermal efficiency compared to the rest which is 17.3%. Concurrently, 0.34MPa pressure of water falls second with only 12.9% efficiency calculated and 0.1MPa pressure of water has the least efficiency with only 10.6% recorded. This may be due to setup imperfection or design error since some leakage in the boiler were suspected.

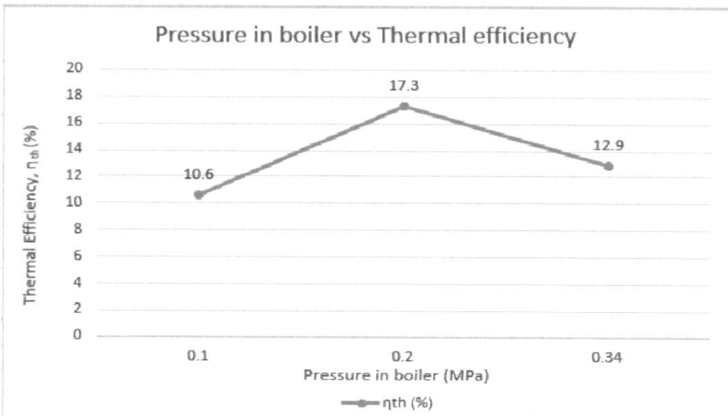

Figure 12: Pressure in the boiler versus thermal efficiency.

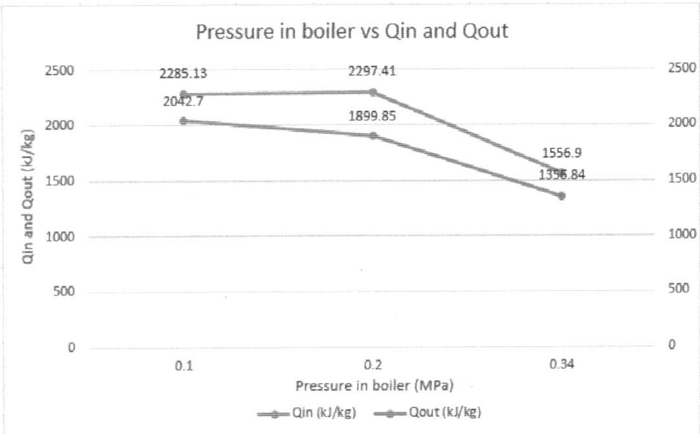

Figure 13: Pressure in the boiler versus energy added in the boiler and energy rejected in the condenser.

4. Conclusion

Experimental setup for Steam Rankine Cycle (SRC) were designed and fabricated. Several experiments were conducted to identify the effect of working fluid volume and boiler pressure on system performance. Results indicates expected trend but with low efficiency due to the design inefficiency. Further improvement in the setup is required to have consistent and improve result for future analysis.

Acknowledgement

The work is funded by the International Islamic University Malaysia (IIUM) Research Initiative Grant Scheme (RIGS). The authors would like to acknowledge the support from the Department of Mechanical Engineering, IIUM for the support to carry out this work.

References

[1] Ali Kilicarslan N M 2005 A comparative study of water as a refrigerant with some current refrigerants International *Journal of Energy Research* 947-959

[2] Bamgbopa M O 2012 Modeling And Performance Evaluation of an Organic Rankine Cycle (ORC) With R245FA as Working Fluid *Master of Science Thesis*

[3] Nouman J 2012 Comparative Studies and Analyses of Working Fluids for Organic Rankine Cycles – ORC *Master of Science Thesis*

[4] Yamamoto T T et al. 2001 Design and Testing of the Organic Rankine Cycle *Energy* vol 26 issue 3 239-251

[5] Matsuda K 2013 *Chemical Engineering Transactions* Retrieved from Low Heat Power Generation System: http://www.aidic.it/cet/13/35/037.pdf

[6] James A Mathias J R 2009 Experimental Testing of Gerotor and Scroll Expanders Used in and Energetic Modeling of an Organic Rankine Cycle *Journal of Energy Resources Technology* 131(1)

[7] Patel F J 2013 Component Matching for Ultra-Low Grade Thermal *Master of Science Thesis*

[8] Electropedia 2005 *Battery and Energy Technologies* Retrieved from Steam Turbine Electricity Generation Plants: http://www.mpoweruk.com/steam_turbines.htm